HYDROGEOMORPHOLOGY

NIPA® GENX ELECTRONIC RESOURCES & SOLUTIONS P. LTD.
New Delhi-110 034

www.nipaers.com
Browse, Search, Read & Buy...
Print Books
eChapters
eBooks
Publishers
Forthcoming
Subject Catalogues
New Titles
Online Resources on:
✓ Current Affairs
✓ Reasoning, Logic & Aptitude
✓ Competitive Examinations
✓ English Language Lab
✓ Writing & Pronunciation Tools
✓ Personality Development
✓ Online Programmes for Professionals
✓ Interview Preparation
Pay Using
PhonePe
Google Pay
paytm
PayPal
UPI
CC Avenue

HYDROGEOMORPHOLOGY

FUNDAMENTALS APPLICATIONS AND TECHNIQUES

Dr Md. Babar
Reader, Geology Department
Dnyanopasak College
Parbhani (M.S.)

NIPA® GENX ELECTRONIC RESOURCES & SOLUTIONS P. LTD.
New Delhi-110 034

NIPA® GENX ELECTRONIC RESOURCES & SOLUTIONS P. LTD.

101,103, Vikas Surya Plaza, CU Block
L.S.C. Market, Pitam Pura, New Delhi-110 034
Ph : +91 11 27341616, 27341717, 27341718
E-mail: newindiapublishingagency@gmail.com
web: www.nipabooks.com

For customer assistance, please contact
Phone: + 91-11-27 34 17 17
Fax: + 91-11- 27 34 16 16
E-Mail: feedbacks@nipabooks.com

ISBN: 978-81-19215-15-7

Composed and Designed by NIPA®.

Preface

Hydrogeomorphology is the science relating to the geographical, geological and hydrological aspects of water bodies and changes to these in response to flow variations and to natural and human caused events such as heavy rainfall or channel straightening. The knowledge of hydrogeomorphology is important for planning and management activities concerned with the surface of the earth. The resource managers and planners for agricultural landuse need detailed, timely, accurate and reliable data on the extent, location and quality of land and water resources and climate characteristics. The hydrogeomorphological processes comprise the physical and chemical interaction between the earth's surface and the natural processes acting upon it to produce landforms in association with their hydrological characteristics.

This book is intended to serve as a guide to those who are concerned with various facets of hydrogeological and geomorphological studies – students, academic and the professional in diverse fields like geology, hydrogeology, geomorphology, irrigation and agricultural engineering and environmental management concerns. Others who will find the book useful are administrators, policy makers, economists whose interest in assessment studies are concerned with formation and evaluation of plans for development and management of the natural recourses. I have endeavored to present the readers with diverse backgrounds the basic principles, various methods and techniques of hydrogeomorphological analysis of the area *i.e.* the basin or watershed. Special emphasis has been laid on aspects dealing with morphometric analysis of the basin, geologic, structural and geomorphological control on the occurrence and movement of the groundwater.

The whole book is divided into 12 chapters.

The *Chapter 1* gives the introduction and definition of hydrogeomorphology in association with the aspects of hydrogeological units, geology, climate and surface processes and groundwater movement. The chapter also discusses the hydrogeomorphology in practice and status of interdisciplinary research and relations in hydrogeomorphology. *Chapter 2* is dealing with the general aspects of mechanism and processes, basic tools, methods of data collections, properties of earth materials, scales of explanation and dimensions of hydrogeomorphological units along with the hydrological cycle, water balance and water budget. The *Chapter 3* deals, firstly with surface water resources including the streams and rivers, lakes and ponds and reservoirs and secondly with groundwater resources with reference to physical properties, chemical characteristics, quality parameters and biological characteristics. *Chapter 4* deals with the classification of aquifers in to confined unconfined and perched aquifers along with the study of aquifer properties like porosity and permeability, specific yield and specific retention, groundwater flow and Darcy's law. The *Chapter 5* presents the hydrological properties of igneous, sedimentary, metamorphic and unconsolidated rocks. *Chapter 6* includes the hydrogeomorphological characters of uniclinal structures and hydrogeomorphic expressions of folds, faults, joints and lineaments.

Chapter 7 seeks to analyse the hydrogeomorphic characteristics of types of drainage pattern and morphometric attributes including methods of stream orders, bifurcation ratio, length ratio, area ratio, basin shape or basin configuration, drainage density, stream frequency, slope and altrimetric analysis. All these morphometric attributes are explained with reference to the research work carried by the author. The aim of *Chapter 8* is to deal with the hydrogeomorphological studies of erosional, depositional, denudational, and structural landforms. The applications of hydrogeomorphic studies are discussed in *Chapter 9* with reference to the groundwater potential zones, groundwater management, water resources assessment, watershed development, artificial recharge studies and hydrogeomorphic assessment method for wetlands.

The *Chapter 10* presents various techniques in hydrogeomorphology including conventional techniques, remote sensing, aerial photo-interpretation, geographical information system, global positioning system, digital elevation model, groundwater information system, digital surface terrain modelling, multi-criteria evaluation, thermal infra red mapping and microwave data and side-looking airborne radar. The *Chapter 11* deals with the guidelines for hydro-geomorphological mapping, generation of thematic maps geomorphological map, landuse map, soil map, groundwater potential map, digital map and hydrogeomorphological action plan map. The *Chapter 12* is concerned with the Social – Cultural – Administrative consideration besides the education and training in hydrogeology for professionals, technicians, farmers, women and masses.

Confronted within the limitations and the tedious task as compared to the time and other busy assignments, the author has every right to confess at the outset that he might have fallen pray to some omissions or some prejudices, which could have introduced discrepancies in the treatment and sometimes even in outcomes and would have denied the work a sort of perfection, for which he has a genuine craze for all of which he would like to be excused.

For bringing out this book I humbly submit my feelings of gratitude towards venerated Dr. R.D. Kaplay, School of Earth Sciences, S.R.T.M. University, Nanded, for his consistent guidance, keen interest, thought provoking discussion, informal and cordial attitude and whole hearted support during the entire work. I am also indebted to Dr. P.L. More (Principal), Dr. P.B. Khanale (Vice-Principal) and Dr. R.B. Muley (Head, Department of Geology), Dnyanopasak College, Parbhani, who gave all the possible help and inspiration for preparation of the book. I would like to express my indebtedness for the affection and blessings from my family members, which had been and will remain a constant source of inspiration throughout my life.

Dr Md. Babar

Contents

Chapter 1

Introduction

⇒ This chapter gives the introduction and definition of Hydrogeomorphology; it also deals with the aspects of hydrogeological units, geology, climate and surface processes and groundwater movement. The chapter also discusses the hydrogeomorphology in practice and status of interdisciplinary research and relations in hydrogeomorphology.

The term hydrogeomorphology can be divided into three terms hydro- means water including both surface and groundwater; geo- means the earth and morphology- is the surface expression of the features in the form of landforms. This means that the hydrogeomorphology is dealing with the aspects of water, rocks and earth's morphological features (land). Of these water and land are most important natural resources for human beings.

Water is synonymous with life. This is a natural resource and an absolute necessity for the survival of the living beings. Human beings have an organic relationship with water. For living and livelihood, human beings solely depend upon the availability of water. They have an inalienable right on water. This is a community resource and cannot be made a tradable commodity, violating normative law of nature. Water is among the most precious gifts of nature to mankind. Most of the ancient towns of the world grew on the banks of the river, and some of the ancient civilizations are known by the names of rivers. Rivers have always been important sources of drinking water and the means of sustaining human, animal and plant life. Whenever one is dealing with water resources development and utilization, it has to be kept in mind that not only quantity of water but also quality of water is very important. The well being and development of our society is dependent on the availability of water.

Groundwater is a most important natural resource required for drinking, irrigation and industrialization. The resource can be optimally used and sustained only when quantity and quality of groundwater is assessed. It has been observed that lack of standardization of methodology in estimating the groundwater and improper tools for handling the same, leads to miscalculation of estimation of groundwater. It is essential to maintain a proper balance between the groundwater quantity and its exploitation. Otherwise it leads to large-scale decline of groundwater levels, which ultimately cause a serious problem for sustainable agricultural production. A possible solution for such problems is micro level planning, and use of standard methodology for assessing the groundwater. In recent years micro level planning has gained acceptance, since it can be locally applied and readily managed by self-sufficient rural governance.

Rationalization issues have been main research tasks in hydrology especially over the past decade (Adyalkar *et al*, 1996, Babar 1999, Bedi and Bhan 1978, Beven and Kirkby 1979, Bloschl and Sivapalan, 1995; Gupta *et al,* 1986, Kulkarni 1994, Singh 1990 and Srinivasa Rao *et al,* 1997). Recently, scaling problems have become more relevant through the need for hydrologic computer models simulating the water balance of large areas. These models are useful in, for example, flood forecasting or improving atmospheric circulation models. Large-scale models cannot incorporate detailed and physically based description of processes, because of unknown boundary conditions and limited computing capacities. Parameterization of boundary conditions and simplification of models are therefore two necessary steps towards the development of hydrologic models for larger scales. One research objective within this broad range of topics is the definition of parameters describing effects of landform structure and topology on hydrologic processes, which we term 'effective geomorphometric parameters'.

On a qualitative basis, it is well known that hydrologic processes are influenced by geomorphometric properties like local slope angle, convergences or drainage density (Gregory and Walling 1973, Rodriguez-Iturbe, and Valdes 1979 and Sharma, Amin 1996). There exist some approaches quantifying these relations through drainage basin parameters (Babar 1998, Balasundaram 1977, Bhattacharya 1996

Kale *et al,* 1994, Moore *et al,* 1991and Singh Savindra 1981) and model conceptions, like the geomorphic instantaneous unit hydrograph (Bloschl and Sivapalan 1995; Moore *et al,* 1991 and Tarboton 1997). However, a general quantification of these effects is still a research task. Recent advances in the analysis of landform morphometry through the availability of high resolution Digital Elevation Models (DEMs), powerful Geographic Information System (GIS) techniques (Lawrence 1985, Tarboton 1997 and Tarboton and Rodiriguez-Iturbe 1991) and remote sensing applications (Babar 2001, Babar 2002a, Babar 2002b, Sharma and Jugran 1992 and Tiwari and Rai 1996) enhance the research efforts, which are discussed in the *Chapter No. 10.*

Definition of Hydrogeomorphology

The term 'Hydrogeomorphology' designates the study of landforms caused by the action of water (Scheidegger, 1973). By this definition almost all of geomorphology is Hydrogeomorphology, because water is one of the most important agent in forming and shaping of landforms. From the groundwater point of view integration of geological, structural and hydrogeological data with hydrogeomorphological data is very much useful in finding out the groundwater potential zones with fruitful results. The science relating to the geographical, geological, and hydrological aspects of water bodies, and to changes to these aspects in response to low variations and to natural and humancaused events, such as heavy rainfall or channel straightening is the hydrogeomorphology.

Knowledge of landuse and hydrogeomorphology is important for planning and management activities concerned with the surface of the earth. The resource managers and planners for agricultural landuse need detailed, timely, accurate and reliable data on the extent, location and quality of land and water resources and climate characteristics. The data on landuse potential and the conservation needs can help in planning for uses that will maintain the quality of land. The application of satellite remote sensing for land use surveys and mapping is gaining importance largely because of its ability to provide rapid and reliable data within a given time framework.

Hydrological Units

A hydrological unit is defined as "an area of land, above or upstream from a specific point on a stream, which is defined by a hydrological boundary that includes all the source areas that could contribute surface water runoff, directly or indirectly to the designated outlet points" (NRCS, 1995). The term hydrological unit is synonymous to the basin, river basin, drainage basin, catchment basin, river catchment or the watershed. Of these most commonly used terms are the basin, catchment and watershed.

The hydrological unit can be classified on the basis of size, shape, sediment yield, drainage density and other land characteristics. As far as dimensions of hydrological units are concerned, it could be as small as a rooftop or a parking lot measuring few square meters and as large as the Amazon River Basin, covering an area of 705 million hectares. The requirement of size of hydrological unit varies with the objectives and approach of the agency involved in its management. In multidisciplinary approach to resource management, an agricultural engineer would select the hydrological unit measuring few hundred hectares to plan intensive soil conservation and water harvesting measures; similarly hydrogeologists might locate observation wells for groundwater monitoring in a fairly large hydrological unit; a meteorologist would prefer to cover a further large hydrological unit to record the weather data and a hydrologist engaged in gauging the river discharge and flood forecasting would target the entire river basin covering an area of few million hectares (Khan *et al,* 2001). Despite the multidisciplinary and multi-jurisdictional approach, the effect of each activity is complimentary over the other, *e.g.,* if the rainwater is conserved *in situ* and the excess runoff is harvested at feasible points by an agricultural engineer, the impact is observed as an increased groundwater table by the hydrogeologists and helps an agronomist to prepare water budget for different landuse systems. The overall effect of different activities results in harmonizing the floods, drought and the ecosystem of the river basin.

Development programme concerning optimum utilization of natural resources are now increasingly oriented with watershed as an integral unit. A watershed is a natural entity conforming to the increasing homogeneity of geomorphic sculpturing process. Watershed management implies rational utilization of land and water resources for optimal and sustained production with the minimum hazard to natural resources and environment. It requires collection and analysis of great deal of information on physical relationship of vegetation-soil-water to land management to ensure economic and social progress of a watershed. The success of planning for developmental activities depends on the quality and quantity of information available on both natural and socio-economic resources. Therefore, accurate and reliable data base generation and management is extremely important for devising ways for optimal planning and management of watersheds.

Now discussing about hydrological units we know that geomorphologists and hydrologists often view streams as being part of drainage basin. A drainage basin is the topographic region from which a stream receives runoff, through flow and groundwater flow. Drainage basins are divided into number of hydrological units, which are separated from each other by topographic barriers called a watershed (Fig. 1.1). A watershed represents all of the stream tributaries that flow to some location along the stream channel. The number, size, and shape of the drainage basins found in an area varies with the scale of examination. Drainage basins are arbitrarily defined based on the topographic information available on a map. The quality of this information decreases as map scale becomes smaller.

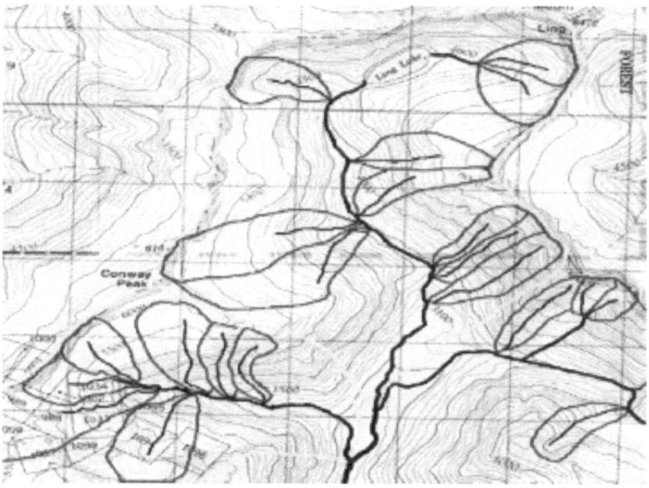

Fig. 1.1. The drainage basins as determined from a topographic map sheet showing the first order drainage basin.

Hydrologists commonly view drainage basins as open systems. Inputs to these systems include precipitation, snowmelt and sediments. Drainage basins lose water and sediment through evaporation, deposition and stream flow. A number of factors influence input, output, and transport of sediment and water in a drainage basin. Such factors include topography, soil type, bedrock type, climate, and vegetation cover. These factors also influence the nature of the pattern of stream channels.

Geology

Geology is considered as an important natural resource since it regulates genesis formation and type of soil, occurrence, distribution and flow of groundwater, and determines presence of minerals. Geology deals with different rock types that constitute the earth's crust, its origin, composition, chronology and tectonic settings. The geological surveys are based on field observations made along a traverse line at regular intervals on a topographic base.

In India, the Geological Survey of India (GSI), (Government of India) has been carrying out geological mapping since last 140 years. The results of the geological mapping are published by GSI in the form of reports and maps.

These maps depict the various geological features / lithological units, their litho succession, characteristics of bedding, foliation, trends, setting and fractures in the form of faults, *etc*. The geological setting as depicted in the maps facilitates the evaluation of terrain condition from the point of view of disaster management, for example, since most of the earthquake prone areas coincide with the fault zones, the geological and structural maps prepared by the GSI will be useful for construction of a geodynamic framework of the regions and also for vulnerability analysis. Even the hydrogeomorphological studies also depend up on the geological setting of the area.

Climate and Surface Processes

Rainfall is a major weather parameter and is a precious natural resource. It is received in measurable and sometimes predictable amount. Similarly, of all the resource factors, climate has the most direct significance, on the general characteristic and broad pattern of biological activity. The broad climatic pattern can be regarded as a permanent feature, although the land use/land cover to some extent can modify the microclimate.

The rainfall and the climate data for the areas are to be collected from the Meteorological Departments. The spatial and temporal distribution of rainfall needs to be depicted for the region as a whole in form of isohyets. For plotting of Theissen Polygon Weighted rainfall of the regions and number of sunshine days/sunshine hours must be known. The average monthly rainfall (mm) and average number of rainy days for minimum of past 10 years should be shown on the map and also depicted in a tabular form as a part of the text. The available mean monthly temperatures (averaged for past 30 years), relative humidity (%) and wind velocity and direction are to be presented in the form of bar charts for any other suitable representation as well as tables.

Groundwater Movement

Groundwater is located beneath the surface of the earth and fills cracks and pores in rock formations. The upper boundary of groundwater is the water table. The zone of aeration (or the vadose zone) is known as the unsaturated zone. It occurs in the top layers of the soil where not all pore spaces are filled with water – (some are filled with air, others with water and some only partially filled with water). During rainfall, water infiltrates downwards through the unsaturated zone and may remain there, or in the case of high rainfall, excess water will reach the water table as discussed in *Chapter 4*.

Pore spaces and cracks found below the water table are completely saturated with water. Groundwater occurs in this zone of saturation. Water beneath the surface is called subsurface water, but water found in the saturated zone is called groundwater. The distance from the ground surface to the water table varies from place to place, but generally, the water table is deeper beneath hills and shallower beneath valleys.

Groundwater moves very slowly, usually at rates of a few feet per month or year (Fetter 1994, Karanth 1989, Lohman 1972 and Todd 1995). Rates of a foot per month are considered fast. Flow is continuously occurring, but rates of flow vary within the aquifer with deeper water moving more slowly. The direction of flow is the point of recharge to the point of discharge. Because of the slow rate of movement and long travel distances from recharge to discharge, the groundwater may be tens of years or more between recharge and discharge areas (Fetter 1994, Karanth 1989, Lohman 1972 and Todd 1995).

In nature the groundwater movement is three-dimensional. The three variables influenced by the time factor are:

(a) Elevation of water level;
(b) Rate of groundwater movement;
(c) Percentage of saturation of aquifer spaces.

Silent features of groundwater movement in Deccan traps are listed below:

- Groundwater movement in Deccan trap is of unsteady nature.
- The groundwater movement appears vertical when the aquifer is half pierced and lateral when it acquires slope and space through the aquifer.
- Upward movement indicates full saturation or an upward movement on regional scale indicates the reinforcement to the water table.
- Hydraulic cross sections are useful for knowing the distribution of energy potential causing groundwater movement in Deccan traps.
- In the groundwater regime of Deccan traps the groundwater movement is a process inherent with the stratigraphic disposition of lava flows.
- Step like rise in elevation of lava flows on surface and termination of individual flows in subsurface provides the energy potentials along with the fluid potential in connected drainage lines for groundwater movement from one place to another. In this situation the influence of time factor is supreme.

Hydrogeomorphology in practice

Knowledge of landuse and hydrogeomorphology is important for planning and management activities concerned with the surface of the earth. The resource managers and planners for agricultural landuse need detailed, timely, accurate and reliable data on the extent, location and quality of land and water resources and climate characteristics. The data on landuse potential and the conservation needs can help in planning for uses that will maintain the quality of land. The application of satellite remote sensing for landuse surveys and mapping is gaining importance largely because of its ability to provide rapid and reliable data within a given time framework.

Development programmes concerning optimum utilization of natural resources are now increasingly oriented with hydrological unit as an integral unit. As a part of hydrological unit the watershed is a natural entity conforming to the increasing homogeneity of geomorphic sculpturing process. Watershed management implies rational utilization of land and water resources for optimal and sustained production with the minimum hazard to natural resources and environment. It requires collection and analysis of great deal of information on physical relationship of vegetation-soil-water to land management to ensure economic and social progress of a watershed. The success of planning for developmental activities depends on the quality and quantity of information available on both natural and socio-economic resources. Therefore, accurate and reliable data base generation and management is extremely important for devising ways for optimal planning and management of watersheds.

Interdisciplinary research and relations

One of the driving forces for the development of applied studies has been the need for hydrogeomorphology to re-examine its position at the interface of the earth, environment, engineering, agriculture, social and economic sciences. Many authors discussed the changing roles and relations of physical and human environment, the nature of the subdivisions within hydrogeomorphology and the relative position of hydrogeomorphology with respect to geomorphology and to other disciplines (Babb 1997, Brinson 1993, Brown 1980, Chorley 1971, Clayton 1980, Dury 1972 Hare 1973, Jones 1976, Jones 1983, Scheidegger 1973, Taylor 1981, Ward 1978 and Worsley 1979).

Present is the era of interdisciplinary research and relation of almost all the sciences with each other. It is obvious that hydrogeomorphological applications are in harmony with the growing awareness for human benefit and with the public and political interests in environmental management problems. Hydrogeomorphologists are now required to have an understanding of the socio-economic and technical planning of their surveys so that they are fully compatible with the needs of the social science disciplines and especially those of rural and urban planners (Agrwal 1995, Akili and Fletcher 1978, Betz 1975, McGill 1964 and Patil *et al* 1999).

Jones (1983) in transactions argued that, there is clear area, the man-environment arena where the geosciences and the socio-economic political sciences overlap and there is an urgent need for research on what he defined as:

i) Natural Hazard Impact – to change planning and management practices and influence policy formulation.
ii) Environmental Auditing – to assess changes in natural environment systems and evaluate the need for new practices.
iii) Resource Assessment – to provide a basis for development.
iv) Impact Assessment – to predict future changes (includes environmental impact statements and impact avoidance assessment).
v) Ex-post Audits or Ex-post Assessment – to review in hindsight the accuracy of predication and the success of projects and policies.

Such work, all of an applied nature, clearly states one possible future for hydrogeomorphology and indeed the earth sciences as a whole in which it strives to re-establish its links with the 'human' sciences. Moreover it has the close association with the other sciences including geology, geophysics, sedimentology, ecology, soil science, pedogeomorphology, Quaternary geology and with the various branches of civil engineering, soil mechanics, rock mechanics and hydrology.

It is also worth recording that the tender documents for many international development and environmental projects issued by various agencies such as DoE, UNDP and World Bank; now include clauses requiring an understanding of the hydrogeomorphology of a development site. The documents further require that the knowledge of landforms, hydrogeology materials and earth surface process be utilized in the design of structures, remedial works, planning frameworks, land zoning plans, building ordinance and code of practice. Hydrogeomorphological knowledge is now being utilized as an individual input to the planning, development and construction process with a consequent improvement to the success of the project and the wise use of the earth as the home of man. This is a most powerful argument for developing interdisciplinary applied research since it suggests that hydrogeomorphology is adapting to the needs of society and growing in status and responsibility.

Chapter 2

Fundamentals of Hydrogeomorphology

⇒ This chapter is dealing with the general aspects of mechanism and processes, basic tools, methods of data collections, properties of earth materials, scales of explanation, dimensions of hydrogeomorphological units, hydrological cycle, water balance and water budget.

Present day hydrogeomorphological research is the explanation of landscape and landforms: what they are, how they function and have developed with reference to the hydrological conditions. Research is a much more complex undertaking than giving names to landforms or even the study of processes, both in terms of what we mean by process and how explanations are approached. It is worthwhile examining some reasons for this, as it shows the increasing complexity of the subject and lead to ideas of how hydrogeomorphology can be studied. Here are some of the fundamentals required for hydrogeomorphological studies:

Mechanism and Processes

'Mechanism' provides explanations in detail by describing physical or chemical effects. 'Process' signifies the integration of mechanisms in time. Thus, for example when we talk about 'fluvial process' or, more specifically, about the 'mechanism of saltation', then we have to discuss about the energy source and processes of carrying the material. A process can thus be viewed as the simultaneous operation of a set of specific mechanisms over a specific time period.

When discussing geomorphological processes we come across different terminologies such as fluvial, glacial, aeolian processes, *etc.* This assumes that the landforms produced by these agencies are essentially distinct from one another because they look different. However, in hydrogeomorphological studies the landforms are studied with reference to the groundwater condition of the area. In discussing a landform whose origins are in question, the morphometric criteria supported with climatic criteria and hydrogeological characteristic are considered.

In recent year there has been a tendency of geomorphologists and hydrologists to converge with sedimentologists and actually look at the materials of which many landforms are composed (Acreman, and Sinclair 1986, Babar and Kaplay 1999a, Brinson 1993, Hammond 1954, Iqbaluddin *et al,* 1997 - King 1950, Leeder 1997, Mertes *et al* 1996, Nakamura 1986 and Subramanyan and Mani 1981). The sedimentologists investigate sediments by examining grain size, shape, etc as well as their internal structure. Thus the studies of mechanisms help to improve explanation and add to clues from morphology and distribution.

Basic tools required

Geomatics Tools

The broad objectives for the use of geomatics tools for are listed below:

- Design a digital geographic database identify spatial and attribute data requirements with their sources
- Create a digital geographic database as per the design integrating both spatial and attribute data from multiple sources
- Carry out satellite image interpretation for the region
- Identify, design and carry out geographic analysis
- Modeling of digital data to identify hot spots and find answers for what-if scenarios
- Generate hardcopy map outputs on paper media

Geographic Database Design

The design extent of the geographic database should be based on the following broad guidelines:

- The spatial extent of the geographic database must cover the total basin falling under the states of Bihar and West Bengal.

- All spatial data spanning entire area shall be at a map scale of 1:250,000.
- Maps of smaller regions within area, such as maps of hydrogeology, geomorphology and landuse pattern shall be at a scale 1: 250,00.
- Data from multiple sources at different scales is to be integrated in the geographic database.
- All maps spanning entire area shall be geo-coded, *i.e.* provided with latitude/longitude values.
- The minimum administrative unit at which demographic and socioeconomic data is to be integrated shall be a block (Tehsil).

Based on the above guidelines the geographic database of area can be designed at two levels:

- *Macro database* comprising maps that span the entire area; all maps in the macro database can be at a scale of 1:250,000; all maps in the macro database are geo-coded
- *Micro database* comprising detailed maps in the micro database depend on the source data and range from 1:3000 to 1:250,000: the maps in the micro database must be geo-coded.

Data collection

The data for hydrogeomorphic analysis include the topographical maps, geological maps of the area, soil map, rainfall and climatic distribution map, geomorphological map, population density map and groundwater fluctuation map. The sources for spatial data in the geographic database in India are in the form of printed maps from the following agencies/publications.

- Survey of India for toposheet maps.
- National Bureau of Census for population density map.
- Geological Survey of India for geological maps.
- Indian Meteorological Division for climate and rainfall distribution data maps.
- National Bureau of Soil Survey and Landuse Planning, Nagpur for soil maps.
- All India Soil and Land Use Bureau for landuse and land cover map.

- Satellite data can be collected from various national and international remote sensing agencies and space research organizations. In India the satellite data can be acquired from National Remote Sensing Agency, (NRSA), Hyderabad, and Indian Institute of Remote Sensing (IIRS), Dehradoon. The data of IRS 1B, 1C, 1D and LANDSAT imageries can be acquired from these agencies.

In addition to the above sources of map data, specially prepared maps from following organizations can be used.

- Central Mining Research Institute (CMRI), Dhanbad
- National Environmental Engineering Research Institute (NEERI), Nagpur
- Centre for Interdisciplinary Studies of Mountain and Hill Environment, University of Delhi, New Delhi.
- Maharashtra Remote Sensing Application Centre (MRSAC), Nagpur.
- Central Groundwater Board (CGWB), Kolkata.
- Water and Land Management Institute (WALMI), Aurangabad.
- Groundwater Survey and Development Agencies of different states.

Properties of earth materials

Hydrogeomorphological knowledge can help alleviate many environmental problems concerned with soil erosion, flooding, draught as well as natural hazards of many kinds. There can be little doubt that, as hydrogeomorphologists learn more about the behaviour of materials they should be better fitted both to understand landforms and converse with engineers and conservationists. There must be simplified approach in the hydrogeomorphology for better understanding of the mechanism and processes. These will be helpful in finding out the nature of both material and the agencies which produce geomorphic features have not changed over geological time *i.e.* the laws of physics and chemistry have not changed.

Materials studied by the hydrogeomorphologists are those upon which processes may act to produce landforms in association with hydrological conditions. Traditionally, rock, weathered material, superficial deposits and soils have been regarded as those materials, which are of basic interest to the hydrogeomorphologists. Initially, the study of the influence of materials upon landform and upon landscapes was achieved largely qualitatively by structural geomorphology but, it has been appreciated that geomorphologists needs to be more aware of the physical characteristic of materials, often expressed in a quantitative way, as a basis for understanding the behaviour of particular materials when they are subjected to the influence of specific processes. The improved understanding of earth material was advocated by Whalley (1976) when he proposed a landform-materials science which would require a knowledge of their physical properties including shear strength, porosity and mineral composition.

Rock, the material of the solid earth, is an aggregate of mineral particles, the composition of which varies according to primary (formation) processes and secondary (diagenetic) processes. Rocks may be classified into three broad groups on the basis of their formation, namely igneous, sedimentary and metamorphic.

Scales of explanation

With respect to the significance of geomorphometric properties in hydrology, scaling effects have to be considered, meaning that (1) runoff-morphometry relations, which tend to be invariant over certain spatial ranges and (2) spatial thresholds affecting changes in these relations have to be determined (Bloschel and Sivapalan, 1995 and Wood, 1995). Translating scale properties from one discipline to the other is an open research question (Anderson and Burt, 1990).

In general, local scale, hill slope scale and catchment scale are often used to distinguish different spatial scales in hydrology (Kirkby, 1988). On the local scale water flow path geometries, flow velocities and quantities are influenced directly by parameters like slope angle and upslope drainage area. The hill slope scale is dominated by runoff production mechanisms influenced by soil properties (partitioning of overland flow and subsurface flow) and hill slope form (Fig. 2.1). Extracting of typical 'hill slope stripes' has been one strategy to represent hill slope hydrology within larger scale catchments. On the basin scale the hydrograph is influenced by basin morphometry, which can be expressed by representative attributes for catchment height distribution (relief indices), length and form of the basin (form indices) and parameters describing the drainage network (Cooke and Doornkamp, 1990; Gregory and Walling, 1973; Schemidt and Dikau communicated). It is also well known that meso-scale or macro-scale landform types affect hydrologic characteristics significantly.

Regarding these scaling effects, different research methodologies have to be applied in the study of the problem. GIS provide sets of useful techniques and tools for investigations on different spatial levels (given that sufficient data is available).

Sensitivity of dynamic hydrologic processes varying in time and space to comparatively static boundary conditions, land surface morphometry and soil features requires different methods for modeling, analysis and visualization of data. Therefore an approach integrating several geo-computational concepts, including spatial analysis of different types of data, static modeling of spatial structures, dynamic 4D-modelling of hydrologic processes, and statistical analysis techniques these scales were chosen.

Local scale

The local scale approaches included investigations of the relevance of morphometric control of local hydrologic processes and the utility of soil-morphometry relations in hydrologic modeling (Fig.2.1). The map algebra tools to analyse the spatio-temporal development of overland flow depth resulting from model runs in sub-catchments of the area can be represented by the spatio-temporal scale.

Catchment scale

Relations between drainage basin parameters and hydrologic indices have been investigated in numerous studies.

Cooke and Doornkamp, (1990) and Gregory and Walling, (1973) provide useful discussion of these issues. Quantitative description of scales in basin morphology and landuse and the establishment of functional relationships between these parameters and hydrologic catchment characteristics have been found to be useful for modeling (Gupta *et al.,* 1980 and Rodriguez-Iturbe and Valdes, 1979) and prediction (Acreman and Sinclair, 1986 and Sauer *et al.,* 1983).

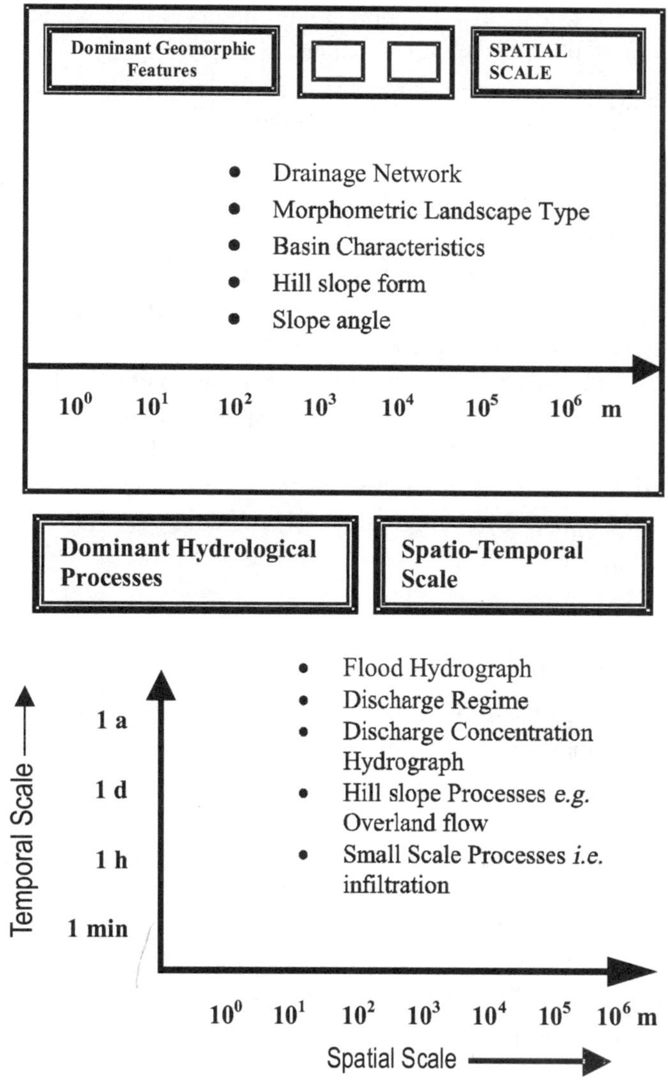

Fig. 2.1. Scales in hydrology and geomorphology. The figure shows in a crude way some dominant features of each discipline in a spatial and spatio-temporal context.

The influence of basin morphometry on runoff in order to identify geomorphometric parameters is the catchment scale and it has hydrologic relevance ('effective parameters'). In order to identify relationships between morphometry and hydrology, the model runs were carried out under the same homogeneous soil, landuse and initial moisture conditions. The catchment parameters can be extracted using a collection of tools, including standard GIS tools, special software and specially developed algorithms. Regression techniques can be applied to get relationships between hydrologic indices and catchment parameters.

Regional scale

A general aim of the regional scale studies is to define meso-scale geomorphometric landform types, reflecting similarities in their hydrologic behaviour. Discriminate analysis using these hydrologic units and representative geomorphometric parameters calculated for the sub-catchments gave evidence of explanatory geomorphometric parameters for the hydrologic classification. On the other hand geomorphometric landform units can be classified by implementing an algorithm proposed by Hammond (1964) and Dikau *et al* (1995). Finally, a hydrologic-morphometric landform classification can be produced from the previous results.

Dimensions of hydrological units

A drainage basin or watershed is a natural unit draining run-off water to a common point. Watershed forms a convenient, clearly defined and an unambiguous topographic unit available on the basis of stream network; the concept of watershed as planning unit for development of land and water resources is widely accepted. In India, since 1974, various developmental projects such as Drought Prone Area Programme (DPAP) Desert Development Programme (DDP), Hill Area Development Programme (HADP), *etc.*, adopt watershed as basic unit for development. Therefore, it is decided to delineate watersheds at various levels of hierarchy based on drainage network.

Watersheds are natural hydrologic entities that cover specific areal extent of land from which rainwater runs to defined gully, stream or river at any particular point. The size of watershed is dependent on the size of the stream, river, the point of interception of stream or river, the drainage density and its distribution. In order to identify and describe a developmental unit on watershed basis, a suitable framework for defining watershed characteristics, identification and codification is necessary. Such a system has already been developed for the country on 1:1 million scales by the All India Soil and land Use Surveys (AIS and LUS), Ministry of Agriculture, Government of India. The system adopted a hierarchical system of approach like water resources regions, basin, catchment, sub-catchment and watershed.

In the AIS and LUS system due of limitations of scale (1:1 million), the smallest unit *i.e.* watershed measures in the range 800-1000 km^2. However, for implementation/operational purposes of developmental activities, it would be desirable that the area of watershed measures around 5-10 km^2. Keeping this scheme in mind, a nine-fold classification has been devised, which is shown in below.

Nomenclature	Scale of Delineation	Area (Km^2)
1. Water Resources Region		550000-600000
2. Basin		95000-100000
3. Catchment	1:100,00,00	30000-40000
4. Sub-Catchment		7000-8000
5. Watershed		800-1000
6. Sub-Watershed	1:250,000	150-300
7. Macro-Watershed	1:50,000	30-50
8. Mini-watershed		10-30
9. Micro-Watershed		5-10

Towards delineation of watershed, first it is preferable to locate the largest unit and identify water divide line for finding out river catchment. The catchment is further divided into number of smaller units (sub-catchment) and sub-catchment to watershed and so on until an areal extent of 5-10 km^2 is identified as micro-watershed.

Survey of India toposheets (1:50,000 scale) in combination with the satellite data (Geocoded FCC print) is to be used for watershed delineation, as the toposheet provides location, drainage network and contour. The satellite data will be useful in updating information on water bodies, drainage, *etc.* Satellite imageries of pre-monsoon, post-monsoon and summer can be used.

For deriving information on surface water body and drainage, transparent polyester film (Garware film 75 micron) containing minimum base details should be used. The major rivers, streams and streamlets are to be carefully extracted. Overlaying the details obtained on the post-monsoon satellite image the extent surface water spread in the tank and ponds can be indicated. This process should be repeated for all seasonal data of the satellite imageries in order to show the correct picture of the surface water spread.

Hydrological cycle, water balance and water budget

Water is the source of all life on the earth. The distribution of water, however, is quite varied; many locations have plenty of it while others have very little. Water exists on earth as a solid (ice), liquid or gas (water vapor). Oceans, rivers, clouds, and rain, all of which contain water, are in a frequent state of change (surface water evaporates, cloud water precipitates, rainfall infiltrates the ground, *etc.*). However, the total amount of the earth's water does not change. The circulation and conservation of earth's water is called the "hydrologic cycle" (Fig.2.2).

Hydrology is defined as the properties, circulation, storage and distribution of water on the earth's surface, as well as beneath the surface, and water within the atmosphere (Fetter 1994, Karanth 1989, Lohman 1972 and Todd 1995). All of this water is part of the hydrologic cycle or the continuous cycle of water through the land and atmosphere by a series of transformations.

Water vapor in the atmosphere comes from water evaporating from streams, lakes, oceans, wetlands and other open water surfaces. Water vapor also comes from evaporation of soils and transpiration by plants. Precipitation results when water vapor in the atmosphere condenses to form clouds. Precipitation may remain on the surface of the ground and evaporate later. The fallen water can also infiltrate the ground surface and percolate to groundwater storage. Overland flow can also occur, which may cause an integrated drainage network that could include lakes and wetlands (Fetter 1994, Lohman 1972, Strahler 1964 and Todd 1995).

Nature recycles the earth's water supply through a process of hydrologic cycle. This cycle operates continuously and receives energy from the sun. The hydrologic cycle consists of inflows, outflows and storage. Inflows add water to the different parts of the hydrologic system, while outflows remove water. Storage is the retention of water by parts of the system. Because water movement is cyclical, an inflow for one part of the system is an outflow for another.

Looking at an aquifer as an example, percolation of water into the ground is an inflow to the aquifer. Discharge of groundwater from the aquifer to a stream is an outflow (also an inflow for the stream). Over time, if inflows to the aquifer are greater than its outflows, the amount of water stored in the aquifer will increase. Conversely, if the inflows to the aquifer are less than the outflows, the amount of water stored decreases. Inflows and outflows can occur naturally or result from human activity.

The hydrologic cycle begins with the evaporation of water from the surface of the ocean. As moist air is lifted, it cools and water vapor condenses to form clouds. Moisture is transported around the globe until it returns to the surface as precipitation. Once the water reaches the ground, one of two processes may occur; (1) some of the water may evaporate back into the atmosphere or (2) the water may penetrate the surface and become groundwater. Groundwater either seeps its way into the oceans, rivers, and streams, or is released back into the atmosphere through transpiration. The balance of water that remains on the earth's surface is runoff, which empties into lakes, rivers and streams and is carried back to the oceans, where the cycle begins again.

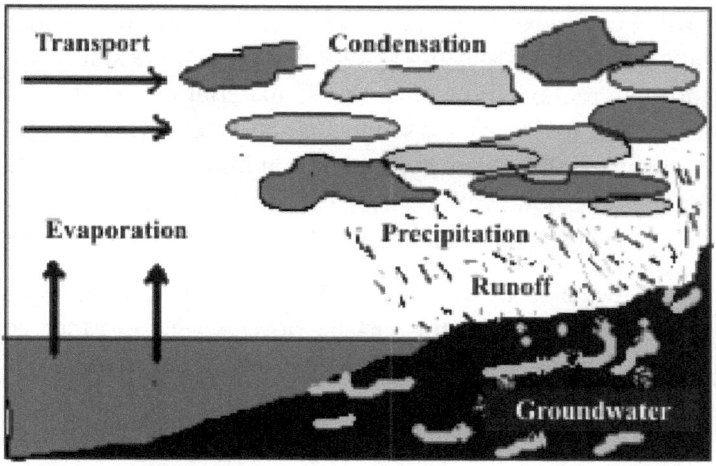

Fig. 2.2. Hydrological Cycle.

By the time these clouds reach the shoreline, they are filled with snowflakes too large to remain suspended in the air and consequently, they fall along the shoreline as precipitation. The intensity of lake effect snowfall can be enhanced by additional lifting due to the topographical features (hills) along the shoreline. Once the snow begins to melt, the water is either absorbed by the ground and becomes groundwater, or goes returns back to the lake as runoff. Evaporation of warm surface water increases the amount of moisture in the colder, drier air flowing immediately above the lake surface. With continued evaporation, water vapor in the cold air condenses to form ice-crystal clouds, which are transported toward shore.

A hydrologic budget considers water availability. A wetland's hydrologic state is based on the three following factors:

1. The balance between the inflows and outflows of water
2. The surface contours of the landscape
3. Subsurface soil, geology and groundwater conditions.

The amount of water entering a wetland over time is about equal to the amount of water leaving the wetland (Fetter 1994, Lohman 1972 and Todd 1995). In the hydrologic budget, inputs are derived from precipitation, surface water inflow and groundwater inflow. Water inputs are the major source of nutrients to wetlands, except in nutrient-poor wetlands such as bogs.

Outputs come from evaporation, combined surface and groundwater outflow and transpiration. These water outflows from wetlands can remove biotic and abiotic material including dissolved organic carbon, toxins, excess sediments and excessive salinity (Mitsch and Gosselink 2000). When hydrologic conditions in wetlands change, the biota may respond with immense changes in species composition, richness and ecosystem productivity (Mitsch and Gosselink 2000).

When water is confined to a channel, surface outflow can be measured with equations for surface flow. A weir or other control structure may be used to estimate the outflow of a wetland.

Chapter 3

Water Resources

≈≈

⇒ This chapter deals with (i) Surface water resources including the streams and rivers, lakes and ponds and reservoirs (ii) groundwater resources with reference to physical properties, chemical characteristics, quality parameters and biological characteristics.

Surface Water Resources

Surface water is a general term describing any water body which is found flowing or standing on the surface, such as streams, rivers, ponds, lakes and reservoirs. Surface water originates from various combinations of sources:

(i) Direct precipitation – rainfall which falls directly in the surface water body;

(ii) Surface runoff – rainfall which has fallen onto the surrounding land and that flows directly over the surface into the water body;

(iii) Interflow – excess soil moisture which is constantly draining into water body; and

(iv) Water table discharge – where there is an aquifer below the water body and the water table is high enough, the water will discharge directly from the aquifer into the water body.

The quality and quantity of surface water depends on a combination of climate and geological factors. The recent pattern of rainfall, for example, is important in enclosed water bodies such as lakes and reservoirs where water is collected over a long period and stored. The water storage in streams and rivers is in a constant stage of movement and depend on the weather conditions.

Streams and Rivers

Generally in rivers the flow is greater in the rainy season as compared to winter and summer in India. Streams and rivers modify the landscape through the movement of water and sediment. Streams are very powerful erosive agents especially during periods of flooding. The sediments removed by streams are usually deposited downstream in floodplains, lakes, and ocean basins. Streams and their floodplains vary in their characteristics along the typical long profile of a stream. The grade of the long profile represents a balance between erosion and deposition processes. At the headwaters as streams flow quickly (because of a steep gradient) on the horizontal strata in the hilly areas the narrow V-shaped valleys are created called gorge and canyon (Fig. 3.1). These V-shaped valleys are also known as saw-cut valleys or slots or gashes. These are quite common features of the major Himalayan Rivers in their mountainous tracts, especially in the inner and the middle Himalayas. In these areas the depositional features are rare.

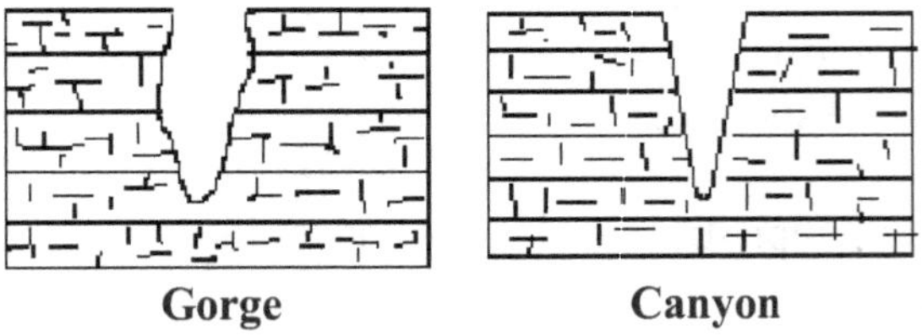

Fig. 3.1: Horizontal strata in the hilly areas produce narrow V-shaped valleys with gorge and canyon.

Further down the profile a profound change occurs in the stream gradient. This change reduces velocity causing the stream to drop its coarser sediments in a floodplain (Fig. 3.2). The channels of these streams are braided. The gradient of the stream becomes very shallow in the last part of the long profile. Now the stream channel takes on meandering and floodplain deposits are spread over a large area (Fig. 3.3).

The amount of water flowing through a stream channel is commonly referred to as stream discharge. Discharge quantity is controlled by water velocity, width of the channel, and the depth of the channel. These variable also change with velocity and increases or decreases over time. Flow within the stream channel is always at a

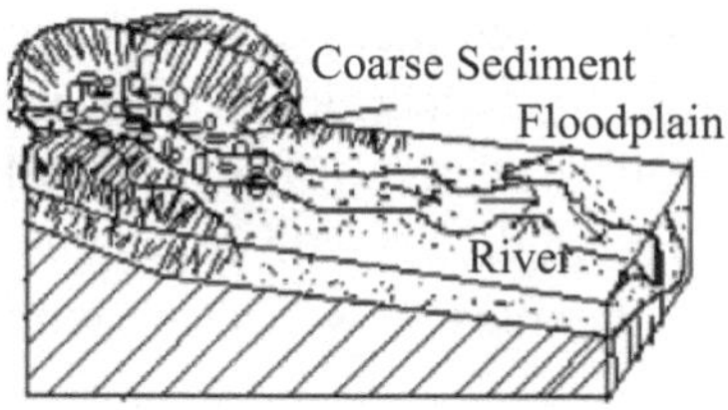

Fig. 3.2: Floodplain in hilly terrain showing coarser sediments.

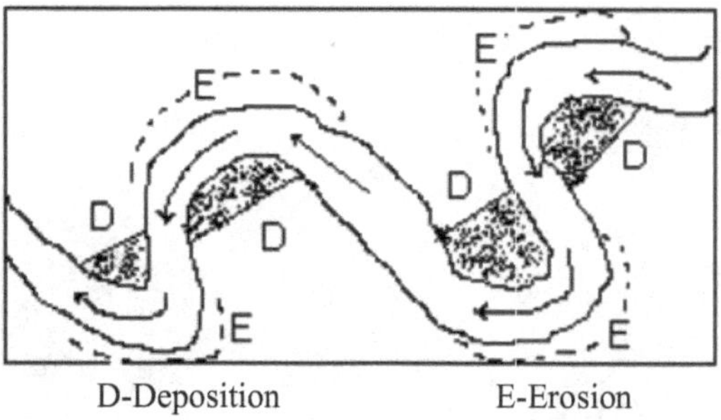

Fig. 3.3: Stream channel showing meandering and floodplain deposits.

maximum at the center of the stream. The channel bottom and banks reduce flow velocity because of friction. In three-dimensions, the zone of maximum flow or thalweg moves from side to side and up and down. A closer examination of flow indicates that three types of flow occur: laminar flow, turbulent flow and helical flow.

All streams carry sediment. The actual amount is related to the characteristics of the materials on which the stream flows. Usually, an increase in discharge results in an increase in the amount of sediment that is carried. The amount of sediment within the stream can be carried in three ways: as the bed load; suspended load and dissolved load (in the form of solution).

One of the most obvious landforms created by streams are the channels they flow in. We can classify stream channels into three types. Channels located in the upper reach of streams are V-shaped, narrow, and deep with little associated floodplain. These channels are often cut into bedrock and limited bed sediments consist loose rocks and boulders. Stream channels become shallow, wide, and braided when the stream gradient changes from being steep to gently sloping. Shallow braided channels are created by the presence of sand and gravel beds in a floodplain (Fig.3.4). These sediments are deposited because a sudden change in gradients causes a reduction in flow velocity. Often a drastic change in grade also causes the deposition of an alluvial fan. Channels become U-shaped and meandering with continued reduction in stream gradient. Meandering channels flow over an extensive floodplain of complex deposits.

Fig. 3.4: Braided channel of Suwarnarekha River (W. Bengal) showing the presence of sand and gravel in the floodplain.

We can describe a number of features within the stream channel. Streams carrying coarse loads develop sand and gravel beds when velocity is reduced spatially or temporally. In straight streams, bar deposits form in relation to the thalweg. Some other features associated with the thalweg include riffles and pools. Some other features common within the stream channel are ripples (Fig.3.5). These features form on the channel bottom when the bed is composed of sand and silt.

Fig.3.5: Channel of Suwarnarekha River (W. Bengal) showing ripples on the floodplain deposits.

Stream flow represents a basin's or watershed's response to climatic conditions, especially precipitation. Ideally, each stream site being monitored should have an active stream flow gauging station. Statistical summaries of historical stream flows and associated return periods from rainfall events are also beneficial to put each monitoring year into context. This information helps to characterize the basin's watershed's response to precipitation and will assist in determining sampling frequency. Water Survey of Canada for *e.g.* has a network of hydrometric stations across Alberta that measure continuous stream flow of selected streams and all major rivers.

Stream *discharge* is the rate of stream flow expressed in volume per unit time. Stream discharge can account for much of the variability in water quality data. Thus, it is very important to collect stream flow data from an established stream gauging station and to establish the relationship between stream *stage* (the height of the water surface, measured by the stream gauge) and stream discharge. For example, a greater amount of phosphorus tends to be delivered during periods of high stream flow. If stream samples are collected randomly without any information on stream flow, the water quality in the stream cannot be fully characterized and the information generated from such a sampling program is limited.

Surface runoff is the main mechanism for transporting non-point source contaminants from the land to surface waters. Runoff is a complex function of many factors including precipitation, infiltration and vegetation. Infiltration depends on soil attributes such as texture, structure and depth of the soil horizons. These parameters determine the soil's water-holding characteristics and tendency to erode. Vegetation plays an important role by intercepting rainfall, retarding runoff and enhancing infiltration. Topographic factors such as slope and surface depressions determine flow velocity and surface storage. Geology controls percolation and groundwater seepage, and the mineral content in surface water and groundwater. Information on these characteristics assists in qualitatively explaining the results of a monitoring program.

Required climatic information includes *rainfall intensities* (*i.e.* the rate at which rain falls), *storm durations* (*i.e.* the length of time of a rainfall event at a certain intensity) and temperature fluctuations.

Data obtained from on-site or off-site stations must capture the variability of climatic conditions in the watershed, including information on both extreme and average conditions. Other important weather variables include *diurnal* (*i.e.* daily) temperature fluctuations, wind velocity and solar radiation. These data are particularly useful in computer models to estimate other meteorological information like evaporation/evapotranspiration and forecast 'what if ' scenarios.

River water is of great importance in the global hydrological cycle and for the supply of water to human beings. This is because the behaviour of individual components in the turnover of water on the Earth depends both on the size of the storage and the dynamics of water movement. The different forms of water in the hydrosphere are fully replenished during the hydrological cycle but at very different rates. For instance, the period for complete recharge of oceanic waters takes about 2500 years, for permafrost and ice some 10,000 years and for deep groundwater and mountainous glaciers some 1500 years (Anonym, 2000). Water storage in lakes is fully replenished over about 17 years and in rivers about 16 days.

Based on water exchange characteristics, two concepts are often used in hydrology and water management to assess the water resources in a region: the static storage component and the renewable waters. The static storage conventionally includes freshwater with a period of complete renewal taking place over many years or decades such as large lakes, groundwater, or glaciers. Intensive use of this component unavoidably results in depleting the storage and has unfavorable consequences. It also disturbs the natural equilibrium established over centuries, whose restoration would require tens or hundreds of years.

Renewable water resources include waters replenished yearly in the process of the water turnover of the earth (Table 3.1). These are mainly runoff from rivers, estimated as the volume per unit of time (m^3/s, km^3/year/*etc.*) and formed either within a specific region or from external sources, including groundwater inflow to a river network. This kind of water resource also includes the yearly renewable upper aquifer groundwater not drained by the river systems. However, on the global scale, these volumes are not large compared with the volume of river runoff and are of importance only for individual specific regions.

Table 3.1: Periods of water resources renewal on the Earth (Anonym 2000).

Water of Hydrosphere	Period of renewal
World Ocean	2500 years
Ground water	1400 years
Polar ice	9700 years
Mountain glaciers	1600 years
Ground ice of the permafrost zone	10000 years
Lakes	17 years
Bogs	5 years
Soil moisture	1 year
Channel network	16 days
Atmospheric moisture	8 days
Biological water	Several hours

In the process of turnover, river runoff is not only recharged quantitatively, its quality is also restored. If only man could suddenly stop contaminating rivers, then with time water could return to its natural purity. Thus, river runoff, representing renewable water resources, is the most important component of the hydrological cycle. It exerts a pronounced effect on the ecology of the earth's surface and on human economic development. It is river runoff that is most widely distributed over the land surface and provides the major volume of water consumption in the world. In practice, it is the value of river runoff that is used to estimate water availability and/or deficit in water resources of the region.

Lakes and Ponds

Most of the lakes have an input and an output and so in some ways they can be considered as very slow flowing rivers. The very long period of time that water remains in the lake or reservoir ensure that the water become cleaner due to bacterial activity removing any organic matter present. In standing water of lakes there is much larger population of algae, which are capable of being supported than in rivers and deep lakes and reservoirs may become thermally stratified particularly, during the summer months.

Thermal stratification is caused by variation density of the water in both lake and reservoirs. Water is densest at 4^0C, when it weighs exactly 1000kg per cubic meter (kg m^{-3}). During the summer the sun heats the surface of the water reducing its density, so that the colder denser water remains at the bottom. As the water continues to heat up and exposed to sunlight it provides a very favourable environment for algae. Normally the various nutrients required by the algae for growth are phosphorous and nitrogen, are not present in large quantities. When excess nutrients are present, due to agricultural run-off for example, then massive algal growth may occur. This massive algal growth known as algal bloom result in vast increase in the quantity of algae in the water, such phenomenon is called eutrophication. The algal bloom in the lake or reservoir is very essential for aquatic animal growth like fishes.

Reservoirs

The importance of reservoirs is likely to increase over time as population, economic activity, and irrigation demands grow. More than 39000 dams higher than 15 m were reported in the world in 1996, although the number of smaller impoundments is several times higher. Agricultural communities and industrial, developed urban areas are heavily dependent on reservoir water supply. In most regions, reservoirs are the single most important component of the regulated water supply system. Most of these structures have been designed to trap sediments continuously with an average usable life of about 100 years of reservoir.

The reservoirs have played a significant role in the India's social and economic progress during the past five decades. Without the dams and reservoirs India would have been a thirsty, hungry dark land ravaged with floods and draughts every year. These reservoirs store precious rainwater to irrigate farmlands, generate electricity, supply drinking water and save land from floods and draughts.

Sedimentation in reservoirs has large consequences for the rivers, the impoundments, and the associated hydraulic structures, but the single most important is the loss of water storage capacity. The loss of water storage capacity caused by sediment accumulation makes reservoirs key non-sustainable components of the water supply systems. The estimated annual lost water storage capacity worldwide is about 1 %, however there is a great range of siltation rates. The geographical, geological, geomorphological, and ecological characteristics of the basin or watershed are significant factors for erosion and sediment sources. The climate and the rivers hydrological regime greatly affect sediment deliver. Human activities and changes in land uses are also significant controls on reservoir siltation. Finally, reservoir management practices can mitigate or increase siltation.

The traditional management of reservoirs does not consider the sustainable use of these structures. Large water flows and measures to mitigate soil erosion are used to span the usable life of a reservoir. Due to the high cost of elimination of sediments, siltation has to be considered as a non-reversible process. A new philosophy for sustainable management of reservoirs has to focus on controlling the sediment accumulation. Water sustainability requires that the traditional concept of limited (< 100 years) reservoir usable life be replaced with a longer-term perspective on sustainable use. A balance between water quantity and quality available to human communities and preservation of biodiversity should be achieved, to protect the environment and conserve resources.

Due to the geographical and climatic diversity of India, the average sediment yield is very much variable, although is higher in the northern Indian basins and lower in the southern Indian basins. Detailed sedimentological studies of reservoirs and watershed provide high-resolution depositional histories to define erosion, transport and sedimentation processes in drainage basins Dhruva Narayana (1987). The influence of anthropogenic, climatic and other geographic factors on sediment transport and delivery can be evaluated within this integrated framework.

In India, Dhruva Narayana (1987), has provided a valuable overview of the downstream impacts of soil conservation in the Himalayan region. He indicates that of India's 328 million ha of land area; approximately 175 million ha are suffering from intense soil erosion. He maintains that the Himalayan and lower Himalayan regions have deteriorated extensively because of deforestation, large-scale road construction, mining, and cultivation on steep slopes. In the northeastern Himalaya alone he attributes the serious degradation of 3 million ha to shifting (slashand-burn) cultivation. Siltation rates in twenty-one Indian

River valley projects were higher by 18.2 percent than the originally projected rates (Table 3.2). He cites the Tehri catchment in the Garhwal Himalaya (total area 7,511 km^2, of which 2,328 km^2, are snow- and ice covered) as producing 14.6 million tonnes of silt annually. The annual silt load from the Karnali catchment is estimated at 75 million m^3, equivalent to a denudation rate of 1.7 mm/yr. Singh and Gupta (1982) have determined that 28.2 tonnes/ha/yr are removed from the entire area of the Indian Himalaya. These are colossal figures and merit careful examination. Nevertheless, these figures also lie within the upper limits of derived denudation rates that have been attributed to geological, or natural causes. This is an extremely important point. It implies that, assuming both sets of figures are within reasonable limits of accuracy, at least at a reconnaissance level.

Table 3.2 Sedimentation rates for major Indian Himalayan reservoirs (after Tejwani, 1986).

Reservoir and Year[1]	Annual Rates of Sedimentation		
	Assumed	**Observed & year of observation**	**% of increase observed**
Sutlej (Bhakra) (1971)	4.29	6.00 (1975)	39 9
		6.20 (1979)	45.0
Beas (1974)	4.29	15.10 (1975)	251.0
		23.59 (1981)	449.9
Ramganga (1974)	4.29	17.30 (1973)	303.3
Pohru[2]		7.41 (1973)	-
Giribata[2]		11.60 (1973)	-
Teesta[2]		98.20 (1973)	-
Gumti[2]		3.56 (1973)	-

[1]Year of completion.

[2]No reservoir constructed.

when considered on the regional scale (as distinct from that of a specific small catchment), the natural processes are so predominant that there is no requirement to seek human intervention as the cause of siltation. In other words, artificial reservoirs will silt up rapidly in this very dynamic region regardless of human influences-negative or positive.

Groundwater Resources

Groundwater is water that has seeped down from the surface and accumulated underground, moving slowly to feed lakes, rivers or the sea. It is present in the pore spaces of almost all rocks and sediments. Rocks and sediments that store and transmit groundwater in usable quantities are called aquifers. Almost all rocks and soils contain spaces (pores). A "porous" rock has pore spaces, which may contain water. In a "permeable" rock the pore spaces are connected, so that water may pass through.

Groundwater accounts for almost all (97%) of the freshwater on Earth (excluding ice caps). The remaining is surface water, in rivers and lakes. Groundwater is important: as drinking water – groundwater is naturally filtered by the rock through which it passes, so it is usually safer and more reliable as an environmental resource – groundwater influences river flow and quality, and sustains fresh surface waters, wetlands and marshes.

Water from beneath the ground has been exploited for domestic use, livestock and irrigation since the earliest times. Although the precise nature of its occurrence was not necessarily understood, successful methods of bringing the water to the surface have been developed and groundwater use has grown consistently ever since. It is, however, common for the dominant role of groundwater in the freshwater part of the hydrological cycle to be overlooked. Groundwater is easily the most important component and constitutes about two thirds of the freshwater resources of the world and, if the polar ice caps and glaciers are not considered, groundwater accounts for nearly all usable freshwater. Even if consideration is further limited to only the most active and accessible groundwater bodies (estimated by Lvovitch (1972) at 4×10^6 km^3) then they constitute 95 per cent of total freshwater. Lakes, swamps, reservoirs and rivers account for 3.5 per cent and soil moisture accounts for only 1.5 per cent (Freeze and Cherry, 1979). The dominant role of groundwater resources is clear and their use and protection is, therefore, of fundamental importance to human life and economic activity.

It is easy for the importance of groundwater in water supplies to be underestimated. It is customary to think of groundwater as being more important in arid or semi-arid areas and surface water as more important in humid areas. However, inventories of groundwater and surface water use reveal the worldwide importance of groundwater. The reasons for this include its convenient availability close to where water is required, its excellent natural quality (which is generally adequate for potable supplies with little or no treatment) and the relatively low capital cost of development. Development in stages, to keep pace with rising demand, is usually more easily achieved for groundwater than for surface water.

In the USA where groundwater is important in all climatic regions, it accounts for about 50 per cent of livestock and irrigation water use, and just below 40 per cent of public water supplies. In rural areas of the USA, 96 per cent of domestic water is supplied from groundwater (Todd, 1980). Some very large cities are totally dependent on groundwater. In Latin America, many of the continent's largest cities, Mexico City, Lima, Buenos Aires and Santiago, obtain a significant proportion of their municipal water supply from groundwater. In the valley of Mexico City, over 1,000 deep wells supply $3{,}200 \times 10^6$ m^3 day^{-1}, which is about 95 per cent of the total supply to a population of nearly 20 million people (Foster *et al.*, 1987). In Europe also, groundwater has always played a major part in water supplies. The proportion of groundwater in drinking water supplies in some European countries as in 1988 (Table 3.3):

Table 3.3: Percentage of groundwater in some European countries (UNEP, 1989).

Name of the country	Percentage	Name of the country	Percentage
Denmark	98%	Netherlands	67%
Portugal	94%	Luxembourg	66%
Germany Fed. Rep.	89%	Sweden	49%
Italy	88%	United Kingdom	35%
Switzerland	75%	Spain	20 %
Belgium	67%	Norway	15%

Many of the major cities of Europe are, therefore, dependent on groundwater. In Africa and Asia, most of the largest cities use surface water, but many millions of people in the rural areas are dependent on groundwater. For many millions more, particularly in sub-Saharan Africa, who do not as yet have any form of improved supply, untreated groundwater supplies from protected wells with hand pumps are likely to be their best solution for many years to come. In India even though, large part is irrigated by the surface stored water in the form of dams, but also the groundwater is used in the rural and urban areas for drinking and irrigation purposes.

Water is drawn from the ground for a variety of uses, principally community water supply, farming (both livestock and irrigated cultivation) and industrial processes. Unlike surface water, groundwater is rarely used *in situ* for non-consumptive purposes such as recreation and fisheries, except occasionally where it comes to the surface as springs. Consequently, groundwater quality assessment is invariably directed towards factors, which may lessen the suitability of pumped groundwater with respect to its potability and use in agriculture and industry.

Two principal features of groundwater bodies distinguish them from surface water bodies. *Firstly*, the relatively slow movement of water through the ground means that residence times in groundwater are generally of the orders of magnitude longer than in surface waters. Once polluted, a groundwater body could remain so for decades, or even for hundreds of years, because the natural processes of flushing are very slow. *Secondly*, there is a considerable degree of physico-chemical and chemical interdependence between the water and the containing material. The word groundwater, without further qualification, is generally understood to mean all the water underground, occupying the voids within geological formations. It follows, therefore, that in dealing with groundwater, the properties of both the ground and the water are important, and there is considerable scope for water quality to be modified by interaction between the two. The scope for such modification is in turn enhanced by the long residence times, which depend on the size and type of the groundwater body (Freeze and Cherry, 1979, Todd, 1980 and Driscoll, 1986). To appreciate the particular difficulties of monitoring groundwater bodies, it is necessary first to identify and to discuss briefly those properties of ground and water that are relevant to the occurrence and movement of groundwater.

Physical Properties:

Physical properties of groundwater are Colour – Temperature – Odour – Taste – density –Turbidity – Electrical Conductivity (Klimentov, 1983 and Trivedy and Goel, 1986).

The *temperature* of groundwater varies greatly with geological structure and geologic evolution of structural units, as well as with physiographic conditions and the regime of recharge. In the permafrost regions the saline water locally have subzero temperature of about –5°C and lower. In middle latitudes the temperature of shallow underground water vary from 5 to 15° C. The areas of hot spring has temperature of about 35-45°C (*e.g.* Unkeshwar hot spring in Nanded district, India has temperature of 37°) and in the areas of recent and active volcanism, as well as in areas where deep crustal waters are in excess of 100°C (geysers of Kamchatka, Japan, Iceland, America, *etc.*) are also known.

The temperature of groundwater changes with time. The temperature of shallow water is the most variable, whereas below the belt of constant yearly temperature, the groundwater temperature increases with depth in accordance with a geothermic slope. The temperature of 7 to 15° C makes drinking water the most tasty and refreshing. The most valuable water for medicinal purpose (*i.e.* for balneology) is that having the temperature of 35-37° C, which is close to the temperature of human body. To utilize such water neither cooling nor heating is required, therefore it is in wide use at the health resorts and holy places in India (*e.g.* Unkeshwar hot spring in Nanded district, India).

Colour of groundwater is dictated by its chemistry and the presence of impurities. Most groundwater samples are colourless. Hard waters are bluish, ferrous salts and hydrogen sulfide colour the water in greenish-blue, organic humic compounds make it yellowish, while suspended mineral particles make it grayish. Watercolour is determined in a glass cylinder 30 to 40 cm high by looking through the water from above. A useful procedure is to compare the water under examination with the distilled water poured into a similar cylinder.

Groundwater usually have no odour, but sometimes it is detectable (Table 3.4) in some cases for example, the hydrogen sulfide imparts the smell of rotten eggs to the water; stagnant water in some wood cased wells often has an objectionable musty odour; shallow groundwater communicating with swamp water have a special marshy odour. It has been found that the water odour is often related to the bacterial decay of organic matter.

Table 3.4: Intensity and characteristics of water odour.

Point	Odour intensity	Odour characteristics
0	No odour	No detectable odour
1	Very faint	Undetectable for a consumer but can be determined in laboratory
2	Faint	Does not attract consumer's attention but detectable when pointed
3	Notable	Easily found and brings about disapproval of water
4	Distinct	Attracts anyone's attention and makes the water objectionable for drinking
5	Very strong	Makes the water non-potable

Drinking water must be odourless. To determine this heating of water up to 40-50^0 C is recommended. Heated water is poured into a bottle until it is half filled, then the bottle's neck is closed with a cork or a finger and the bottle is strongly shaken 3 to 5 times, immediately after the determination is carried out.

Water may have a *taste* or a *smack* brought about by dissolved mineral matter, gases and impurities. Water containing Calcium and Magnesium Bicarbonates or carbonic acid is nice to taste. Abundant organic compounds make the water sweetish; a saltish taste is due to large amounts of Sodium chloride, while a bitter taste is due to the presence of magnesium and sodium sulfates. Iron ions bring about the special rusty taste.

The *density* of water is the ratio of its mass to its volume at a certain temperature. Unit density is that of distilled water at 4^0 C. The density of water depends upon its temperature and an amount of dissolved salts, gases or suspended particles. The groundwater density ranges from 1to 1.4 g/cm^3.

Water density is usually measured by an areometer or a picknometer. If water contains predominantly sodium chloride, the water density is derived from its salinity in Baume degrees. One degree of Baume gravity is equivalent to the density of 1.000 cm^3 of water containing 10 gm of sodium chloride.

Turbidity in groundwater is caused by the presence of insoluble sediments, organisms and organic matter. Turbidity in the water reduces the light penetration. Light penetration, in fact depends partly on the light flux but mainly on the optical properties of water. Usability of water in certain industries and in the artificial recharge purposes may be restricted by turbidity limits. Properly developed wells are usually free from turbidity.

Turbidity of water can be measured by Secchi disc method. A Sacchi disc is a circular disc having metallic disc of 20 cm in diameter; upper surface of which is painted with alternate black and white colour is used to find the light penetration characteristics of turbid water. It has got a weight at the lower face so as to avoid a drift during lowering in water. A string is attached to it for lowering, which is marked in centimeters.

Sacchi disc is lowered into the water with the help of thread tied to hook in the center. Two readings of depth, one at a point when the disc just disappears, note down the depth by marking on the string as (A) and the other reading is taken as (B) at which it reappears. These readings are to be taken in centimeters. The turbidity is directly calculated from Secchi disc transparency value (Kodarkar *et al,* 1988 and Trivedy and Goel, 1986) *i.e.*

$$\text{Light penetration} = \frac{A+B}{2}$$

Where, A= depth at which Sacchi disc disappears.
B= depth at which Sacchi disc reappears.

Electrical conductivity (EC) of the water is due to the fact that water is an electrolytic solution. Electrical conductivity is directly proportional to the amount of dissolve salts. The distilled water does not conduct the electric current. The electrical conductivity denotes the characteristics of a medium to the passage of electricity. Resistivity is the reciprocal of conductivity. In water quality determination, conductivity is defined as the conductance of a cube of one-centimeter side of a substance is expressed in mhos/cm now used as Siemens (S)/cm. When expressed in mhos/cm the conductivity of most of the natural water is less than unity. Hence for convenience, the conductivity is expressed in micro-mhos/cm (μS/cm), which is equal to one million times the values given in mhos/cm.

It is generally measured with the help of a conductivity meter having a conductance cell containing electrodes of platinum coated with Pt. Black or carbon. These electrodes are mounted rigidly and placed parallel at a fixed distance. Conductance when measured between the electrodes having a surface area of 1 cm^2 and placed at a distance of 1cm is called electrical conductivity. The value of conductivity is calculated as follows:

Electrical Conductivity (EC) = observed conductance X cell constant X temperature factor at 25^0 C.

As natural water commonly contains several salts in solution, no definite relation exists between the conductance and dissolved solids. The following relation can express the relation between the conductance and dissolved solids up to the values of 50,000 μS/cm:

$$\text{EC } (\mu\text{S/cm}) = \frac{\text{Dissolved Solids in ppm}}{F}$$

Where the factor *F* has a wide range from 0.5 to 1.

During the determination of electrical conductivity it is essential to note the temperature of the water sample and find out the factor from the (Table 3.4) given below in (Table 3.5) to convert the values at 25^0C.

Table 3.5: Factors for converting the values of conductivity at 25^0C (After Golterman *et al.*, 1978).

^{0}C	Factor	^{0}C	Factor	^{0}C	Factor
32	0.89	22	1.06	12	1.30
31	0.90	21	1.08	11	1.33
30	0.92	20	1.10	10	1.36
29	0.93	19	1.12	9	1.39
28	0.95	18	1.14	8	1.42
27	0.97	17	1.16	7	1.46
26	0.98	16	1.19	6	1.50
25	1.00	15	1.21	5	1.54
24	1.02	14	1.24	4	1.58
23	1.04	13	1.27	3	1.62

The electrical conductivity of groundwater varies from a few tens of μS/cm, in chemically inert rocks in areas of abundant precipitation, to over 1,00,000 μS/cm in some desert brines. The electrical conductivity of oceanic water is about 50,000 μS/cm.

Chemical Characteristics

Since groundwater often occurs in association with geological materials containing soluble minerals, higher concentrations of dissolved salts are normally expected in groundwater relative to surface water. The type and concentration of salts depends on the geological environment and the source and movement of the water. A simple hydrochemical classification divides groundwater into meteoric, connate and juvenile. Meteoric groundwater is derived from rainfall and infiltration within the normal hydrological cycle and is subjected to the type of hydrochemical evolution. Groundwater originating from seawater, which has been entrapped in the pores of marine sediments since their time of deposition is generally referred to as connate water. The term has usually been applied to saline water encountered at great depths in old sedimentary formations. It is now accepted that meteoric groundwater can eventually become equally saline, and that entrapped seawater can become modified and moved from its original place of entrapment. It is doubtful whether groundwater exists that meets the original definition of connate water, and many authors prefer the non-generic term formation water. Connate water is, perhaps, useful to describe groundwater that has been removed from atmospheric circulation for a significant period of geological time. Formation waters are not usually developed for water supplies because of their high salinity. However, they may become involved in the assessment of saline intrusions caused by the over pumping of overlying aquifers.

Juvenile groundwater describes the relatively small amounts of water, which have not previously been involved in the circulating system of the hydrological cycle, but are derived from igneous processes within the earth. However, juvenile groundwater often cannot be distinguished geochemically from meteoric groundwater that has circulated to great depths and become involved in igneous processes. True juvenile waters unmixed with meteoric water are rare and very localised extent, and are not normally associated with the development and assessment of fresh groundwater resources.

The natural chemical quality of groundwater is generally good, but elevated concentrations of a number of constituents can cause problems for water use. The relative abundance of the constituents dissolved in groundwater is given in (Table 3.6) and a summary of the natural sources and range of concentrations of the principal constituents of groundwater is given in (Table 3.7).

Table3.6: Relative abundance of dissolved constituents in groundwater

Major constituents (1.0 to 1,000 mg l^{-1})	Secondary constituents (0.0001 to 0.1mg l^{-1})	Minor constituents (0.01 to 10.0 mg l^{-1})
Sodium	Iron	Arsenic, Bromide
Calcium	Aluminium	Barium, Cadmium
Magnesium	Potassium	Chromium
Bicarbonate	Carbonate	Cobalt, Copper
Sulphate	Nitrate	Iodide, Lead
Chloride	Fluoride	Lithium, Nickel
Silica	Boron	Manganese
	Selenium	Phosphate
		Strontium
		Uranium, Zinc

Source: After Todd, 1980

High iron levels in groundwater are widely reported from developing countries, where they are often an important water quality issue. Consumers may reject untreated groundwater from hand pump supplies if it has a high iron concentration in favour of unprotected surface water sources with low iron levels but which may have gross bacteriological pollution. The situation is made worse in many areas by the corrosion of ferrous well linings and pump components. Obtaining representative samples for iron in groundwater presents particular difficulties because of the transformations brought about by the change in oxidation/reduction status which occurs on lifting the water from the aquifer to ground level.

Table 3.7: Sources and concentrations of natural groundwater components

Component	Natural sources	Concentration in natural water
Dissolved solids	Mineral constituents dissolved in water	Usually < 5,000 mg l^{-1}, but some brines contain as much as 300,000 mg l^{-1}
Nitrate	Atmosphere, legumes, plant debris, animal excrement	Usually < 10 mg l^{-1}
Sodium	Feldspars (albite), clay minerals, evaporites such as halite, NaCl, industrial wastes	Generally < 200 mg l^{-1}; about 10,000 mg l^{-1} in sea water; ~ 25,000 mg l^{-1} in brines
Potassium	Feldspars (orthoclase and microcline), feldspathoids, some micas, clay minerals	Usually < 10 mg l^{-1}, but up to 100 mg l^{-1} in hot springs and 25,000 mg l^{-1} in brines
Calcium	Amphiboles, feldspar, calcite, gypsum, pyroxenes, dolomite, aragonite, and clay minerals	Usually < 100 mg l^{-1}, but brines may contain up to 75,000 mg l^{-1}
Magnesium	Amphibole, olivine, dolomite, pyroxenes, magnesite, clay	Usually < 50 mg l^{-1}; about 1,000 mg l^{-1} in ocean water; brines may be 57,000 mg l^{-1}
Carbonate	Limestone, dolomite	Usually < 10 mg l^{-1}, but can exceed 50 mg l^{-1} in water highly charged with Na
Bicarbonate	Limestone, dolomite	Usually < 500 mg l^{-1}, can exceed 1,000 mg l^{-1} in water highly charged with CO_2
Chloride	Sedimentary rock (evaporites), a little from igneous rocks	Usually < 10 mg l^{-1} in humid areas; up to 1,000 mg l^{-1} in more arid regions; approximately 19,300 mg l^{-1} in sea water and up to 200,000 mg l^{-1} in brines
Sulphate	Oxidation of sulphide ores, gypsum, anhydrite	Usually < 300 mg l^{-1}, except in wells influenced by acid mine drainage; up to 200,000 mg l^{-1} in some brines
Silica	Feldspars, ferromagnesian and clay minerals, amorphous silica, chert and opal	Ranges from 1-30 mg l^{-1} but as much as 100 mg l^{-1} can occur and concentrations may reach 4,000 mg l^{-1} in brines

Contd...

Table 3.7: *Contd...*

Component	Natural sources	Concentration in natural water
Fluoride	Amphiboles (hornblende), apatite, fluorite, mica	Usually < 10 mg l^{-1}, but up to 1,600 mg l^{-1} in brines
Iron	Igneous rocks: amphiboles, dark micas, FeS, FeS_2 and magnetite, Fe_3O_4. Sandstone rocks: oxides, carbonates, sulphides or iron clay minerals	Usually < 0.5 mg l^{-1} in fully aerated water; groundwater with pH < 8 can contain 10 mg l^{-1}; infrequently, 50 mg l^{-1} may be present
Manganese	Metamorphic and sedimentary rocks and mica biotite and amphibole hornblende minerals contain large quantities of Mn	Usually < 0.2 mg l^{-1}; groundwater contains > 10 mg l^{-1}

Source: Modified from Todd, 1980

Water Quality Parameters

Chemical analysis forms the basis of interpretations of the quality of water in relation to source, geology, climate and use. Water being an excellent solvent, it is important to know the geochemistry of the dissolved solid constituent and methods of reporting analytical data. Methods and procedures of chemical analysis of water samples, which strictly belong to the domain of chemistry, are not generally relevant in groundwater studies.

Groundwater quality is the sum of natural and anthropogenic influences. Multi-purpose monitoring may be directed towards a whole range of groundwater quality issues and embrace many variables. Other categories of monitoring may concentrate on a single quality issue such as agricultural pesticides, an industrial spill, saline intrusion, etc. These have been linked to the principal uses to which groundwater is put and to three levels of industrial development, so that the most important current and future groundwater quality issues can be identified. A major subdivision into urban, industrial and agricultural is made, although it is clear that there is considerable overlap between the first two. The most important groundwater quality issues worldwide are described briefly in the following sections.

Variability in water quality

Aquatic ecosystems are dynamic systems and are influenced by many factors. Water quality monitoring to assess non-point source contamination in such ecosystems is particularly difficult because there is a large degree of natural variability to account for. Factors contributing to variability in water quality data include:

- Weather conditions;
- Hydrologic conditions;
- Sampling frequency;
- Spatial variability;
- Temporal variability;
- Monitoring design;
- Land use and land management practices; and
- Analytical error (laboratory and statistical).

Weather conditions: Some water quality parameters may be positively or negatively correlated to weather conditions such as precipitation. For example, runoff from large rainfall events can pick up and deliver more contaminants to a stream than would be transported during dry weather. Conversely, some dissolved contaminants may not be affected by an increase in surface runoff or stream flow.

Hydrologic conditions and sampling frequency: Flow conditions in streams or seasonal influences on a lake affect water quality. To adequately characterize the quality of a water body, sampling must also account for the range of hydrologic conditions. Sampling streams infrequently or only during low flow periods will yield information that does not adequately characterize the stream's overall water quality. Calculation of ***mass loads*** — the total amount of a substance that is transported by a stream — and ***flow-proportionate stream sampling***— taking more samples during high flows and fewer during low flows — are required to describe ambient stream water quality.

Spatial variability: Spatial characteristics like sampling location can also affect variability in water quality. For example, water quality for certain parameters (*e.g.* dissolved oxygen, suspended sediment) may differ depending on whether the sample was taken at the surface or bottom of a stream. Variability may also be introduced if a sampling station was changed between years.

Temporal variability: The time of day (daily) or the time of year (seasonal or annual) when samples are taken can influence the concentration of a particular water quality variable, such as dissolved oxygen.

Monitoring design: The type of sampling (*i.e.* grab or composite samples) is another way variability can be introduced into a sampling program. ***Grab samples*** are discrete samples that are taken at specific sampling times. ***Composite samples*** are samples taken over a period of time (*e.g.* hours or days) or from a variety of locations and combined into one sample.

Land use and land management practices: A long period of record is required to determine if change or level of impact is due to the effects of a particular practice or to natural variability. Consequently, a long-term commitment is recommended to adequately assess water quality.

Analytical error: Laboratory analysis can also introduce a level of variability or error into water quality data. Quality control and quality assurance components, to validate the data, should be part of any water quality-monitoring program. Statistical analysis tests (*e.g.* analysis of variance, *t*-test) can also introduce a certain level of error into the evaluation of the water quality data.

In general, there is a large amount of natural variability in the quality of water resources. To determine statistically significant differences in water quality or to properly characterize the water quality of an aquatic system, data from several years are recommended to account for this natural variability.

To aid in explaining variability in monitoring results, it is imperative to document sampling protocol (*e.g.* grab or composite sample) and describe conditions at the sampling sites (*i.e.* verbal descriptions of weather conditions, the overall state of the stream, *etc.*).

Water quality parameters

pH: The pH is the hydrogen ion concentration, which expresses the intensity of an acidity or alkalinity of the water. The pH value is the negative logarithm of the concentration of hydrogen ions in moles per liter. In pure water, the dissociated molar concentrations of H^+ (hydrogen) and OH^- (hydroxyl) ions are equal, each being 10^{-7} moles per liter, equivalent to a pH of 7. If the H^+ ions exceeds OH^- ions, as in acids solutions, the pH value is less than 7, while the pH is more than 7 if OH^- ions exceeds the H^+ ions, as in basic solutions.

In most natural waters, the pH value is dependent on the carbon dioxide-carbonate-bicarbonate equilibrium. Hussainy (1967) and Singh and Anandh (1996) believed that pH is an ecological factor in controlling the activities and distribution of aquatic flora and fauna. The rise and fall in the values of free CO_2, CO_3 and HCO_3 are directly related to the pH values and any change in the former complex has a direct bearing on maxima and minima of pH, the heavy rainfall by causing the considerable dilution of water and increasing the buffer effect, cause fluctuation in the pH. Roy (1955) suggested that the high pH and alkalinity are associated with high photosynthetic activity.

Total Hardness (TH)

Hardness is mainly due to the presence of bicarbonates of Ca^{++} and Mg^{++} ions. It is an important parameter in detection of water pollution. The presence bicarbonates of Ca^{++} and Mg^{++} seem to cause the temporary hardness, while the permanent hardness is due to the Sulphate and Chloride.

$$TH = 2.497\ Ca + 4.115\ Mg$$

Where TH, Ca and Mg are all measured in ppm (mg/liter)

Calcium and Magnesium showed certain relationship with other parameters. A direct relationship was observed between those factors are Chloride and Sulphate (Rao *et al* 1993). According to Prasad and Manjula (1986) the richness of bicarbonate in water is concurrent with the view that water rich in bicarbonates is usually rich in Calcium.

Calcium is the most common element found in most natural water at levels ranging from 0-100 mg/liter. It contributes to the hardness property of water and taste. The sources of Ca and Mg in natural water are rock, sewage and industrial waste. At higher pH its concentration is decreased due to precipitation as Calcium carbonate ($CaCO_3$). High concentration in water causes lather formation with soap and not desirable in washing, bathing and laundry. Scale formation takes place along with Magnesium in boilers. It coagulates with soap and makes dirty layers on sinks, tubes *etc.*

Magnesium occurs in all water along with calcium but generally its concentration is very much lower than that of calcium. Hardness of water is generally caused due to Magnesium. Its higher concentration imparts an unpleasant taste to the water.

TH values between 150 and 300 mg/liter means the water is hard and the value greater than 300 mg/liter means it is very hard (Todd 1995). High concentration of TH in water may cause kidney stone and heart disease in human (Jain 1996). The maximum permissible limits of water quality for drinking as given by BIS (1991) and WHO (1992) is 600 mg/lit. BIS (1991) fixed the maximum permissible limits for Calcium (Ca) and Magnesium (Mg) in water at 200 mg/lit and 150 mg/lit respectively (Table 3.8).

Table 3.8: Standards for physical and chemical quality of drinking water

Quality Parameter	**W.H.O. International Standards; 1992**		**Bureau Indian Standards; 1991**	
	Highest Desirable	**Maximum Permissible**	**Highest Desirable**	**Maximum Permissible**
PHYSICAL PARAMETERS				
Turbidity (JTU Units)	5	25	10	25
Colour, Hazen unit (on Pt-Co scale)	5	50	5	50
Taste and odour	Unobject tionable	—	Unobjec tionable	—
CHEMICAL PARAMETERS				
pH	7.0 – 8.5	6.5 – 9.2	7.0 – 8.5	6.5 – 9.2
TDS (mg/l)	500	1500	500	1500
TH (mg/l)	100	500	300	600
Ca (mg/l)	75	200	75	200

Contd...

Table 3.8: *Contd...*

Quality Parameter	W.H.O. International Standards 1992		Bureau Indian Standards, 1991	
	Highest Desirable	**Maximum Permissible**	**Highest Desirable**	**Maximum Permissible**
Mg (mg/l)	<30 if SO4 is 250 mg/l, up to 150 mg/l if SO4 <250 mg/l	150	30	100
Iron as Fe (mg/l)	0.05	1.5	0.3	1.0
Manganese as Mn (mg/l)	0.1	1.0	0.1	0.5
Copper as Cu (mg/l)	0.05	1.5	0.05	1.5
Zinc as Zn (mg/l)	5.0	15.0	5.0	15.0
Chloride (mg/l)	200	600	250	1000
Sulphate (mg/l)	200	400	150	400
Phenol (mg/l)	0.001	0.002	0.001	0.002
Fluorides (mg/l)	0.6 – 0.9	0.8 – 1.78	0.6 – 1.2	1.5
Nitrates (mg/l)	—	45	45	No relaxation
OXIC CONSTITUENTS				
Arsenic (mg/l)	—	0.05	0.05	-do-
Mercury (mg/l)	—	0.001	0.001	-do-
Cadmium (mg/l)	—	0.01	0.01	-do-
Chromium (mg/l)	—	—	0.05	-do-
Cyanide as CN (mg/l)	—	0.05	0.05	-do-
Lead (mg/l)	—	0.1	0.1	-do-
Selenium (mg/l)	—	0.01	0.01	-do-
RADIOACTIVITY				
Gross α – emitters (Pc i/l)	—	3	—	—
Gross β – emitters (Pci/l)	—	30	—	—

Total dissolved Solids (TDS)

The maximum dissolved concentrations in the water are the total dissolved solids (TDS). In natural water TDS generally ranges from 10 to 30,000 mg/lit. Verma *et al* (1978), observed that the excess amount of dissolved solid result into high osmotic pressure, which in turn causes unbalance of osmotic regulation and suffocation of the drain water. Reddy and Venkateswarulu (1987) in the River Tungbhadra (India) found TDS range was from 350 to 503 mg/lit. Similarly, Suvarna and Someshwara (1997) found the average seasonal variation of TDS of about 179.68 mg/lit in monsoon, 196.67 mg/lit in winter and 218 mg/lit in summer in Cauvery River (India).

As regards Total Dissolved Solids (TDS), the BIS (1991) maximum permissible limit is 1500 mg/lit for drinking water. High TDS indicates that the water is highly mineralized, which is, in turn unsuitable apart from potability for industrial applications also. The concentrations of dissolved solids over 2000 mg/ lit produce a laxative affect (Dhembre *et al* 1998 and Jain 1996).

Chlorides (Cl)

It is a natural substance present in all potable water and usually occurring in sewage as metallic salt. Its concentration in fresh water is quite low and generally less than that of sulphate and bicarbonates. The principal sources of Chloride are animal organic matter, sewage from drainages and refuse (Qasim and Siddiqui, 1960; Bulusu *et al* 1967 and Rai 1974). It is found that about 8-15 grams of Sodium Chloride (NaCl) is extracted by a person per day. Chloride cannot be removed biologically in treatment of water, as it is highly soluble with cations and does not sediment and precipitate. Rai (1974) reported sharp increase in the chloride concentration at sewage pollution station and recorded low at unpolluted station. Mishra and Yadav (1978) and Kumar (1980) found the rise of chloride due to increased summer temperature and evapo-transpiration. The BIS (1991) maximum permissible limit for Cl in drinking water is 1500 mg/ lit.

Sulphates (SO_4): Natural water contains higher levels of Sulphate contributed from the rock weathering. In addition to this domestic sewage and industrial effluents also add sulphate to aquatic ecosystem and hence high level of sulphate is an indicator of organic pollution. Odorous condition is easily created when water is over loaded with organic waste to the point that oxygen is removed and the SO_4 formed. The SO_4 act as an electron acceptor, which is often used for the breakdown of organic matter and produce H_2S and a rotten egg smell (Welch, 1998). Biological oxidation of sulphur emission attributes sulphate to the water. Sulphate exhibited a direct relationship with dissolved oxygen, whenever the oxygen content was the sulphate also increased. Singh (1992) studies on water quality of major rivers of Pune (India) and found SO4 of 15.5 mg/lit in Mula River, 30.0 mg/lit in Shivganga River and 13.0 mg/ lit in Bhima River.

The Sulphate (SO_4) concentration in natural water ranges from 0.0 to 1000 mg/lit. As per NRC (1977) SO_4 cannot be removed from the range by any of the common treatments. The maximum permissible limit of SO_4 for drinking water given by BIS (1991) is 400 mg/lit. The high concentration of SO_4 may induce diarrhea similarly the laxative effect may occur at lower concentration if Mg is present in the water at an equivalent concentration.

Total Alkalinity (CO_3 + HCO_3)

Total alkalinity is the measure of the capacity of water to neutralize a strong acid. The alkalinity of water is generally imparted by the salts of carbonates, bicarbonates, phosphates, nitrates, silicates along with hydroxyl ion in free state. Alkalinity less than 100 mg/L is desirable for domestic purpose. However, most of the water is rich in carbonates (CO_3) and bicarbonates (HCO_3) with little concentration of other types of alkalinity imparting ions. Alkalinity values provide guidance in applying doses of alum in water treatment.

Singh and Saha (1981) found alkalinity value 60 mg/lit indicating hard water, which makes the fresh water unpalatable.

Biological Characteristics

Groundwater quality can be influenced directly and indirectly by microbiological processes, which can transform both inorganic and organic constituents of groundwater. These biological transformations usually hasten geochemical processes (Chapelle, 1993). Single and multi-celled organisms have become adapted to using the dissolved material and suspended solids in the water and solid matter in the aquifer in their metabolism, and then releasing the metabolic products back into the water (Matthess, 1982). There is practically no geological environment at, or near, the earth's surface where the pH and Eh conditions will not support some form of organic life (Chilton and West, 1992). In addition to groups tolerating extremes of pH and Eh, there are groups of microbes which prefer low temperatures (psychrophiles), others which prefer high temperature (thermophiles) (Ehrlich, 1990), and yet others which are tolerant of high pressures. However, the most biologically favourable environments generally occur in warm, humid conditions.

Microorganisms do not affect the direction of reactions governed by the thermodynamic constraints of the system, but they do affect their rate. Sulphides, for example, can be oxidized without microbial help, but microbial processes can greatly speed up oxidation to the extent that, under optimum moisture and temperature conditions, they become dominant over physical and chemical factors.

All organic compounds can act as potential sources of energy for organisms. Most organisms require oxygen for respiration (aerobic respiration) and the breakdown of organic matter, but when oxygen concentrations are depleted some bacteria can use alternatives, such as nitrate, sulphate and carbon dioxide (anaerobic respiration). Organisms that can live in the presence of oxygen (or without it) are known as facultative anaerobes. In contrast, obligate anaerobes are organisms, which do not like oxygen. The presence or absence of oxygen is, therefore, one of the most important factors affecting microbial activity, but not the only one. For an organism to grow and multiply, nutrients must be supplied in an appropriate mix, which satisfies carbon, energy, nitrogen and mineral requirements (Ehrlich, 1990).

Most microorganisms grow on solid surfaces and, therefore, coat the grains of the soil or aquifer. They attach themselves with extra-cellular polysaccharides, forming a protective biofilm, which can be very difficult to remove. Up to 95 per cent of the bacterial population may be attached in this way rather than being in the groundwater itself. However, transport in the flowing groundwater is also possible. The population density of microorganisms depends on the supply of nutrients and removal of harmful metabolic products (Matthess, 1982). Thus, in general terms, higher rates of groundwater flow supply more nutrients and remove the metabolic products more readily. Microbe populations are largest in the nutrient-rich humic upper parts of the soil, and decline with decreasing nutrient supply and oxygen availability at greater depths. Many sub-surface microbes, however, prefer lower nutrient conditions. In the presence of energy sources, such as organic material, anaerobic microbial activity can take place far below the soil and has been observed at depths of hundreds and even thousands of meters. The depth to which such activity is possible is determined by the nutrient supply and, in addition, pH, Eh, salt content, groundwater temperature and the permeability of the aquifer.

Microbiological activity primarily affects compounds of nitrogen and sulphur, and some of the metals, principally iron and manganese. Sulphate reduction by obligate aerobes is one of the most important biological processes in groundwater. Both nitrifying and denitrifying bacteria affect nitrogen compounds. Reduction of nitrate by denitrifying bacteria occurs in the presence of organic material in anaerobic conditions, leading to the production of nitrite, which is then broken down further to elemental nitrogen. The possibility of enhancing natural denitrification is currently receiving attention in relation to the problem of nitrate in groundwater. Under aerobic conditions, ammonia (which may be produced during the decomposition of organic matter) is oxidized to nitrite and nitrate. Likewise iron can be subjected to either reduction or oxidation, depending on the pH conditions of the groundwater. In favourable microbiological environments, massive growth of iron bacteria can cause clogging of well screens and loss of permeability of aquifer material close to wells, and may require special monitoring and remedial action.

Microorganisms can break down complex organic materials dissolved in groundwater. Under anaerobic conditions, microbial breakdown proceeds either as methane fermentation or by reduction of sulphate and nitrate (Matthess, 1982). Microbial decomposition has been demonstrated for a whole range of organic compounds, including fuel hydrocarbons, chlorinated solvents and pesticides. Under ideal conditions, all organic materials would eventually be converted to the simplest inorganic compounds. In practice, complete breakdown is never reached, and intermediate products of equal or even greater toxicity and persistence may be produced. Nevertheless, the possibility of using indigenous or introduced microbial populations and enhancing the nutritional status of their environment is currently an important area of research in the quest for effective techniques for remediation of contaminated aquifers (NRC, 1993).

A principal microbiological concern in groundwater is the health hazard posed by faecal contamination. Of the four types of pathogens contained in human excreta, only bacteria and viruses are likely to be small enough to be transmitted through the soil and aquifer matrix to groundwater bodies (Lewis *et al.,* 1982). The soil has long been recognised as a most effective defense against groundwater contamination by faecal organisms, and a number of processes combine to remove pathogens from infiltrating water on its way to the water table. Not all soils are equally effective in this respect. In addition, many human activities which can cause groundwater pollution involve the removal of the soil altogether. Bacteriological contamination of groundwater remains a major concern, especially where many dispersed, shallow dug wells or boreholes provide protected but untreated domestic water supplies.

In summary, microbiological processes may influence groundwater quality both positively and negatively. The former include reducing nitrate and sulphate contents of groundwater, and removal of organic pollutants. The latter include the production of hydrogen sulphide and soluble metals, production of gas and biofilm fouling of well screens and distribution pipes.

The bacteriological investigation of any area is very essential because it clearly reveals the quantitative and qualitative picture of pathogens, which transmit the harmful diseases to human being, through the groundwater available adjacent to sewage disposal areas. The presence of pathogenic bacteria in a high rate, may poses many risks to public health. Certain austerity measures should be taken immediately to prevent pollution of the stream and by the stream. Such problems can be solved by taking the preventive measures like training of natural streams or nalas, proper disposal of refuse, monitoring of tap and valves of pipelines, bleaching of drinking water wells, proper sealing of the abandoned wells in city area and awareness for keeping the clean environment. In order to prevent the pollution, indiscriminate dumping of waste and debris, as well as misuse of stream bank by slum dwellers and villagers has to be prevented. The sewage flow in the stream can be controlled by widening the existing sewage drain and by constructing parallel sewage drain.

Chapter 4

Aquifer Properties

⇒ This chapter deals with the classification of aquifers in to Confined Aquifer, Unconfined Aquifer and Perched Aquifers along with the study of aquifer properties like Porosity and Permeability, Specific Yield and Specific Retention, Groundwater flow and Darcy's law.

An aquifer is a geologic formation, a part of a formation, or a group of formations that will yield water. The quantity of ground water that an aquifer will yield to wells depends partly on its thickness, extent, continuity, and homogeneity, and partly on its physical properties of permeability and porosity.

Groundwater is stored in subsurface void spaces below the water table. The geologic material that stores, transmits, and yields groundwater to wells and springs are called *aquifers*. The key word here is yields. To be an aquifer, it must store and transmit water at rates fast enough to supply reasonable amounts to wells and springs. Therefore, not all groundwater is stored in an aquifer. Productive aquifers can be comprised of sand and gravel, sandstone, limestone and dolomite, or basalt flows, whereas fractured igneous or metamorphic rocks are only marginal aquifers.

The classifications of aquifer are given below:

Classification of Aquifers

Three main types of aquifers exist:

- Confined aquifer
- Unconfined aquifer
- Perched aquifer

Confined Aquifer

An impermeable or semi-permeable layer of rock confines the confined aquifers. They are not recharged from above by vertical infiltration. However, they do need to be connected to an unconfined area through which recharge can occur. Rarely do the confining impermeable layers form a complete barrier to groundwater movement (Karanth, 1989, Freeze and Cherry, 1979 and Todd, 1995). A confined aquifer usually contains clay or silt-sized sediment, a tightly cemented rock, or a mixture of sediment sizes such as found in glacial till. The confining layer inhibits the vertical movement of water into or out of the aquifer. The degree to which that water movement is slowed depends on the thickness and composition of the confining layer (Fig. 4.1).

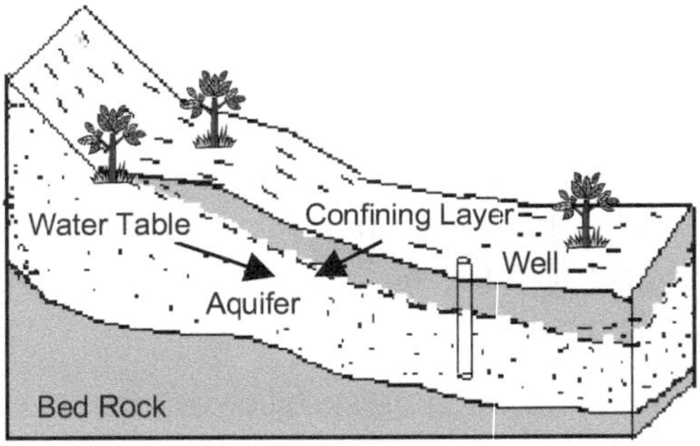

Fig. 4.1: Confined Aquifer.

Confined aquifers are also distinguished from unconfined aquifers, perched water table aquifers, by a "pressure head." Pressure head is a measure of the elevation to which water levels will rise above the top of the aquifer. If the pressure head is greater than the elevation of the ground surface, the well will flow without pumping. This is sometimes referred to as artesian flow from an artesian well.

Water pumped from a confined aquifer moves by three principal processes. These are leakage into the aquifer, expansion of water within the aquifer because of pressure head reduction, and compaction of the aquifer much like squeezing a sponge. The amount of water yielded by each process depends on the degree to which the aquifer is confined and the compressibility of the aquifer. Voids or pore spaces in confined aquifers remain saturated.

Regional precipitation is the source of water recharging confined aquifers. A small portion of this precipitation infiltrates the surface of the ground and is pulled downward by gravity. Downward movement continues until a point called the water table is reached. Below the water table all pore spaces are filled with water. If the surface of this saturated zone is at an elevation greater than the pressure head in a lower confined aquifer, leakage of water into the confined aquifer will continue until the two are in equilibrium. This saturated zone could be an adjacent aquifer, a region in the overlying confining layer, an aquifer above the confined aquifer or a combination of the above. Recharge could also come from an adjacent confined aquifer with a higher-pressure head.

A *confined aquifer* is overlain by a *confining layer or aquitard,* which is geologic material with little or no *permeability/hydraulic conductivity.* This layer does not allow water to pass through or the rate of movement is extremely slow. Often clays and silts, or rocks such as shale comprise confining layers. When a well is drilled into a confined aquifer, the water level in this well may rise higher than the top of the aquifer due to pressure. This well is called an *artesian well* (Fig.4.2). The level to which the water rises is called the *potentiometric surface* (similar to the *water table* in unconfined aquifers).

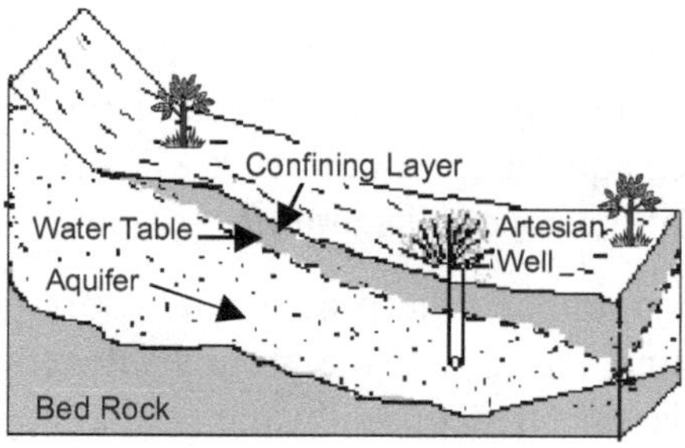

Fig. 4.2: Artesian Well.

A confined aquifer is sandwiched between confining beds (layers of impermeable materials such as clay which impede the movement of water and out of the aquifer). Because of the confining beds, ground water in these aquifers is under high pressure. Because of the high pressure, the water level in a well will rise to a level *higher than* the water level at the top of the aquifer. The water level in the well is referred to as the potentiometric surface or pressure surface.

Even in a confined aquifer, water seeks its own level. Geological strata are not perfectly horizontal. At some point the lithological unit that comprises the confined aquifer is exposed to the surface. The "confined" aquifer is actually unconfined at the recharge zone. In order for pressure to build, the water level in the recharge zone must be at a higher elevation than the base of the confining unit. When a well is drilled through the confining unit, usually far from the recharge zone, the water in this well will rise to the level of the water at the recharge zone. In some instances this may be above the surface of the ground, in which case the well is called a flowing artesian well. This same situation, where the level of the water at the recharge zone is above the base of the confining unit, leads to the appearance of springs or seeps where the confining unit is penetrated by a hillside.

It is important to note that confining beds not only serve to hamper the movement of water into and out of the aquifer, they also serve as a barrier to the flow of contaminants from overlying unconfined aquifers. For this same reason, however, contaminants that reach a confined aquifer through a poorly constructed well or through natural seepage can be extremely difficult and expensive to remove.

Unconfined Aquifer

An *unconfined aquifer* has no confining layers between the *zone of saturation* and the land surface. If a fully screened well is drilled into an unconfined aquifer, the water level will rise to the *water table*, which is the top of the zone of saturation.

In unconfined aquifer, the groundwater only partially fills the aquifer and the upper surface of the groundwater (the water table) is free to rise and decline. The groundwater is at atmospheric pressure. The height of the water table will be the same as the water level in a well constructed in that unconfined aquifer. The water table typically mimics, in a subdued way, the topography of the land surface above, resulting in a water table with hills, valleys, or flat areas. It is important to note that unconfined aquifers, especially those close to the surface, can be vulnerable to contamination from activities on the land surface.

Unconfined aquifers are located in areas where there are permeable geologic formations (Fig. 4.3). The upper surface of this aquifer is the

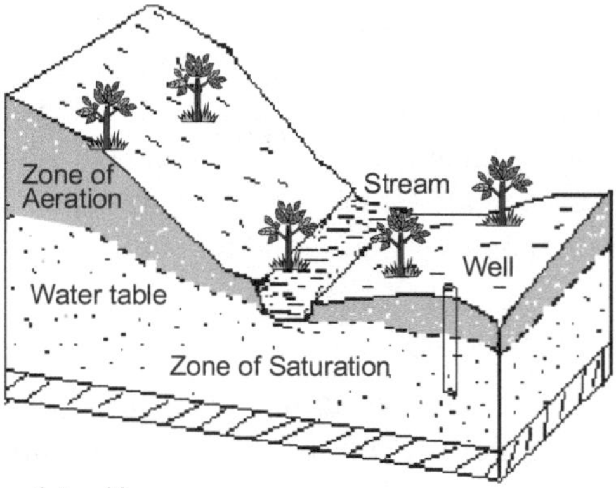

Fig. 4.3: Unconfined Aquifer.

water table, where the rock formations are saturated. These are known as water table aquifers and they receive recharge directly from the infiltration of surface water and rainfall.

Perched Aquifer

Perched aquifers are aquifers that have a confining layer below the groundwater, and sits above the main water table. A *perched aquifer* is a saturated zone within the zone of aeration that overlies a confining layer (Fig. 4.4). A perched aquifer sits above the main water table. A well may be drilled into a perched aquifer, but it usually only yields enough water for a household. A *perched aquifer* is a saturated zone within the zone of aeration that overlies a confining layer. A perched aquifer sits above the main water table. A well may be drilled into a perched aquifer, but it usually only yields enough water.

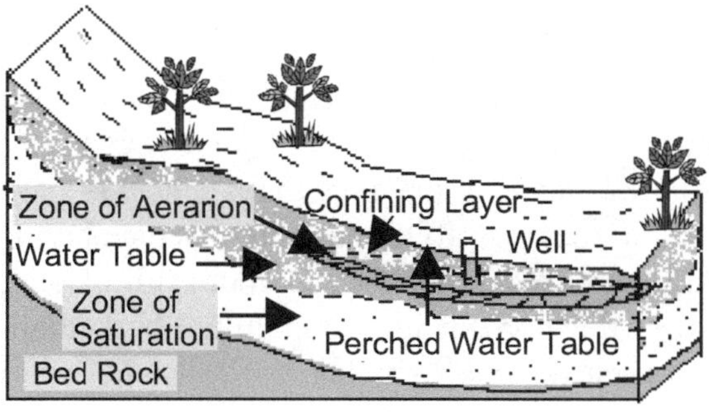

Fig. 4.4: Perched Water Table.

Perched aquifers are located where the groundwater is perched above unsaturated rock formations resulting from a discontinuous impermeable layer (Karanth, 1989, Lohman 1972 and Todd, 1995). They occur in sedimentary formations where weathered layers and ancient soils have created somewhat impermeable layers. Glacial sediments are also common places to find perched aquifers.

Porosity and Permeability

Porosity: The shape and arrangement of soil particles help to determine *porosity*. Porosity or pore space is the amount of air space or void space between soil particles. *Infiltration*, groundwater movement, and storage occur in these void spaces. The porosity of soil or geologic materials is the ratio of the volume of pore space in a unit of material to the total volume of material. A mathematical equation of porosity looks like this: Porosity or $n=V_{void} / V_{total}$. Porosity is often expressed as a percentage of rock or soil void of material, so multiply the answer by 100.

The arrangement or packing of the soil particles plays a role in porosity. In the diagrams to the left, the particles stacked directly on top of each other (cubic packing) have higher porosity than the particles in a pyramid shape sitting on top of two other particles (rhombohedral packing). Can you see the difference in pore space?

What could happen when smaller particles are mixed with larger particles? As the diagram shows, the smaller particles could fill in the void spaces between the larger particles, which would result in a lower porosity.

Not all particles are spheres or round. Particles exist in many shapes and these shapes pack in a variety of ways that may increase or decrease porosity. Generally, a mixture of grain sizes and shapes, results in lower porosity. One important point to remember is that the diameter size of the grain does not affect porosity. Remember, porosity is a ratio of void space to total volume. A room full of ping-pong balls would have the same porosity as a room full of basketballs, as long as the packing or arrangement is similar.

Sands have large pore spaces, whereas clays have many small pore spaces. Both sand and clay can have high porosity. According to

Holt, the sand-plain province has a porosity value between 32% and 38%.

Table 4.1: Porosity Ranges for Sediments

Material	Porosity (%)
Well-sorted sand or gravel	25-50
Sand and gravel, mixed	20-35
Glacial till	10-20
Silt	35-50
Clay	33-60

(C.W. Fetter, 1994)

Permeability: The size of pore space and interconnectivity of the spaces help determine permeability, so shape and arrangement of grains play a role. *Permeability* is a measure of a soil's or rock's ability to transmit a fluid, usually water. Often the term *hydraulic conductivity* is used when discussing groundwater and aquifer properties. Hydraulic conductivity simply assumes that water is the fluid moving through a soil or rock type.

Water can permeate between granular void or pore spaces, and fractures between rocks. The larger the pore space, the more is the permeable of the material. However, the more poorly sorted a sample (mixed grain sizes), the lower the permeability because the smaller grains fill the openings created by the larger grains. "The most rapid water and air movement is in sands and strongly aggregated soils, whose aggregates act like sand grains and pack to form many large pores".

The field coefficient of permeability (P) is defined as the number of gallons of water that will move in 1 day, at the prevailing water temperature, through a vertical section of the aquifer 1 foot square under a *hydraulic gradient* of 100 percent or 1 foot per foot. Coefficients of permeability of less than 100 gallons a day per square foot are considered low, coefficients of 100 to 1,000 are medium, and those of more than 1,000 are considered high. The coefficient of transmissibility (T) is equal to the field coefficient of permeability multiplied by the thickness (m) of the aquifer.

Groundwater is not usually static but flows through the rock. The ease with which water can flow through a rock mass depends on a combination of the size of the pores and the degree to which they are inter-connected. This is the permeability of the rock. Materials, which permit water to pass through them easily, are said to be permeable and those which permit water to pass only with difficulty, or not at all, are described as impermeable. A layer of rock that is sufficiently porous to store water and permeable enough to transmit water in quantities that can be economically exploited is called an aquifer. Groundwater flow may take place through the spaces between the grains or through fissures (Fig. 4.5), or by a combination of the two in, for example, a jointed sandstone or limestone. For any aquifer, distinguishing whether inter-granular or fissure flow predominates is fundamental to understanding the hydrogeology and to designing monitoring systems, particularly for point source pollution incidents.

Permeability describes how easily water passes through unconsolidated sediments (not cemented to form solid rock). If pore spaces are connected and large enough for water to move through the spaces, the rock is permeable. High porosity does not always mean that the rock is highly permeable. Solid bedrock may not have high porosity (doesn't hold a lot of water), but fractures in the rock can make it permeable. On the other hand, clay soils are porous but there are very small pore spaces makes them relatively impermeable. A good source of groundwater supply must be permeable and porous or have permeable layers, which are connected to porous layers where groundwater is stored (Fetter, 1994, Lohman 1972 and Todd, 1995).

Specific Yield and Specific Retention

The coefficient of storage (S) is defined as the volume of water, measured in cubic feet, released from storage in each column of the aquifer having a base 1 foot square and a height equal to the thickness of the aquifer, when the water table or other piezometric surface is lowered 1 foot. In watertable aquifers the coefficient of storage for long periods of pumping is approximately the specific yield and has a range from about 0.1 to 0.3. For artesian aquifers the coefficient of storage generally is very small, ranging from about 10^{-5} to 10^{-3}.

Specific yield is the ratio of volume of water that drains from a saturated rock due to gravity to the total volume of rock. A mathematical equation of specific yield looks like this: Specific Yield or Sy = $(V_{drained} / V_{total})$ X 100. Not all the water stored in pore spaces becomes part of flowing or moving groundwater. Just as water clings to a glass, water also clings to soil particles due to surface tension, cohesion, or adhesion. It forms a thin film around a particle. Thus *specific yield* is always less than porosity.

Unlike porosity, specific yield is influenced by grain size. For example, if two soil samples have the same porosity, but different grain sizes (*e.g.* clay and sand), the sample with smaller grain sizes will have a lower specific yield. Clay has a greater surface area than sand; therefore, more water will remain behind clinging to the clay particle surfaces (Table 4.2).

According to Holt, the sand-plain province has a specific yield between 19% and 30%, which indicates that recharge from *precipitation* should pass easily from the surface to the *zone of saturation*.

Table 4.2: Specific yield of different geological formations.

Material	**Specific Yield (%)**		
	Material	**Maximum**	**Minimum**
Coarse gravel	26	12	22
Medium gravel	26	13	23
Fine gravel	35	21	25
Gravelly sand	35	20	25
Coarse sand	35	20	27
Medium sand	32	15	26
Fine sand	28	10	21
Silt	19	3	18
Sandy clay	12	3	7
Clay	5	0	2

(After C.W. Fetter, 1994)

The coefficients of transmissibility and of storage are used in making quantitative estimates of water available in an aquifer, and of the waterlevel decline that will result from pumping. Controlled aquifer (or pumping) tests can be made to obtain the data required to determine these coefficients.

Drawdown in a well is the lowering of the water table or piezometric surface caused by pumping or artesian flow. The specific capacity of a well is the discharge expressed as rate of yield per unit of drawdown, generally gallons per minute per foot of drawdown.

Groundwater occurs in many different geological formations. Nearly all rocks in the upper part of the Earth's crust, whatever their type, origin or age, possess openings called pores or voids. In unconsolidated, granular materials the voids are the spaces between the grains (Fig. 4.5a), which may become reduced by compaction and cementation (Figure 4.5d). In consolidated rocks, the only voids may be the fractures or fissures, which are generally restricted but may be enlarged by solution (Fig. 4.5 e, f). The volume of water contained in the rock depends on the percentage of these openings or pores in a given volume of the rock, which is termed the porosity of the rock. More pore spaces result in higher porosity and more stored water. Typical porosity ranges for common geological materials are shown in table 4.1.

Only a part of the water contained in the fully saturated pores can be abstracted and used. Under the influence of gravity when, for example, the water level falls, part of the water drains from the pores and part remains held by surface tension and molecular effects. The ratio of the volume of water that will drain under gravity from an initially saturated rock mass to the total volume of that rock (including the enclosed water) is defined as the specific yield of the material, and is usually expressed as a percentage. Typical values are shown in table 4.3.

When rain falls, some infiltrates into the soil. Part of this moisture is taken up by the roots of plants and some moves deeper under the influence of gravity. In the rock nearest to the ground surface, the empty spaces are partly filled with water and partly with air. This is known as the unsaturated or vadose zone and can vary in depth from nothing to tens of meters. In the unsaturated zone, soil, air and water are in contact and may react with each other. Downward water movement in the unsaturated zone is slow, less than 10 m a^{-1} and often less than 1 m a^{-1}.

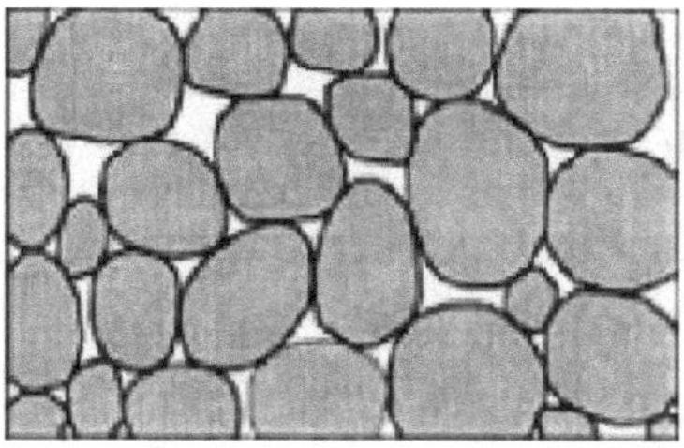

(a) Well sorted sedimentary Deposit with high porosity

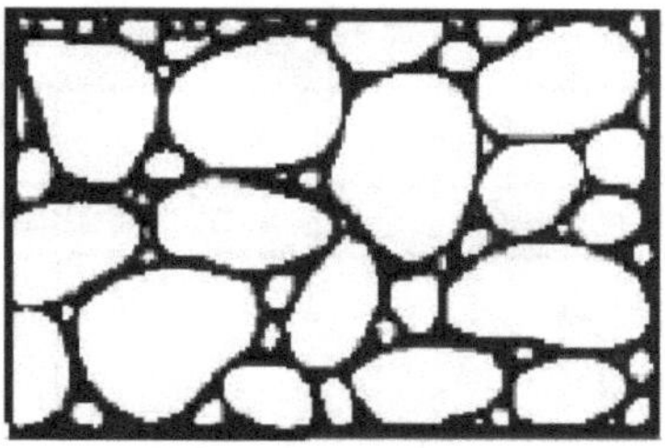

(b) Poorly sorted sedimentary deposit with low porosity

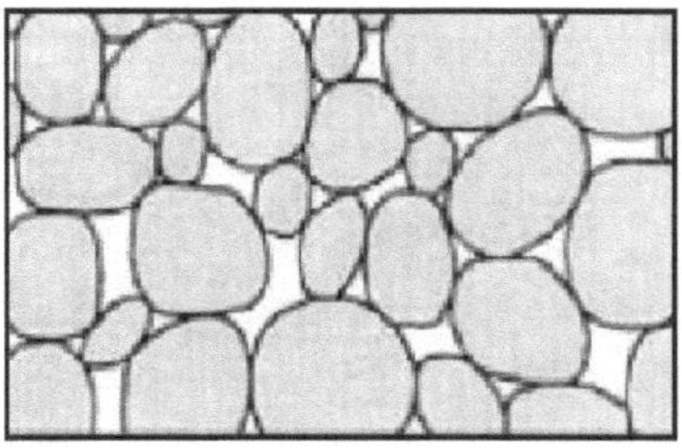

(c) Well sorted sedimentary deposit of porous pebbles resulting in high overall porosity

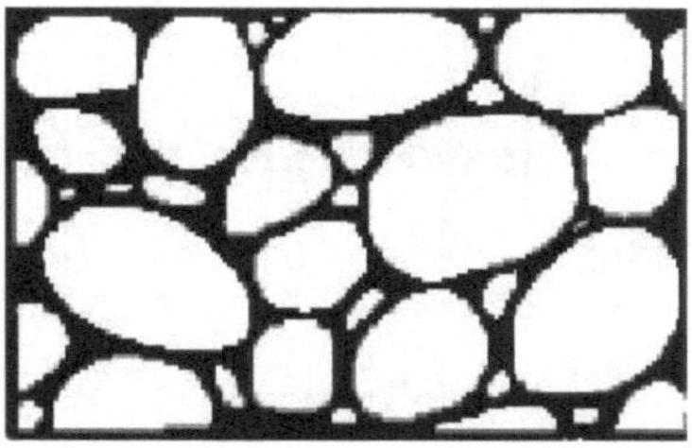

(d) Well sorted sedimentary deposit in which porosity has been reduced by cement deposited between grains

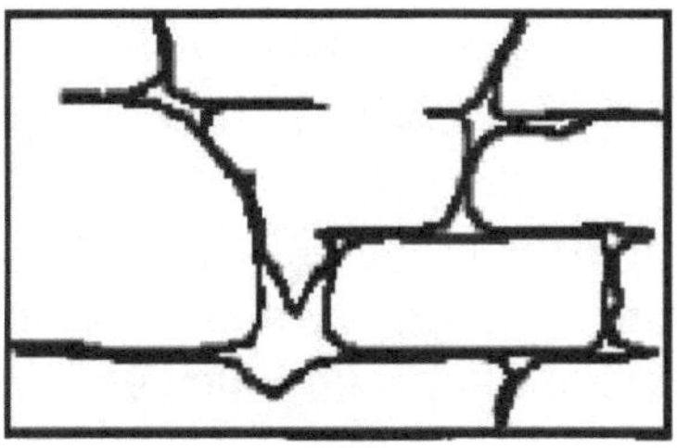

(e) Consolidated rock rendered porous by solution

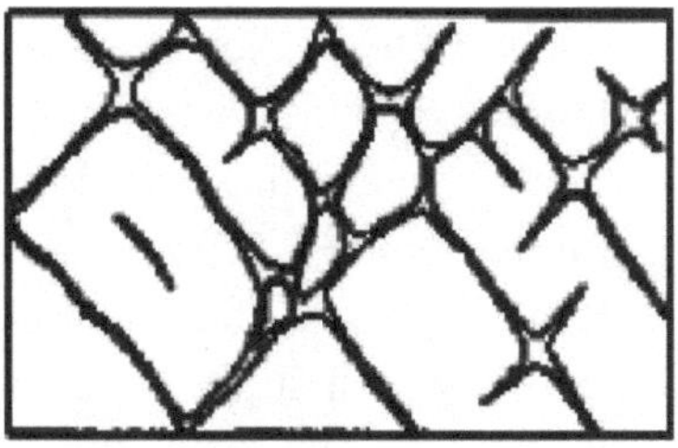

(f) Consolidated rock rendered porous by fracturing

Fig. 4.5: Texture and porosity of typical aquifer materials (After Todd, 1980).

Table 4.3: Porosity and specific yield of geological materials

Material	Porosity (%)	Specific yield (%)
Unconsolidated sediments		
Gravel	25-35	15-30
Sand	25-45	10-30
Silt	35-50	5-10
Clay	45-55	1-5
Sand and gravel	20-30	10-20
Glacial till	20-30	5-15
Consolidated rocks		
Sandstone	5-30	3-15
Limestone and dolomite	1-20	0.5-10
Karst limestone	5-30	2-15
Shale	1-10	0.5-5
Vesicular basalt	10-40	5-15
Fractured basalt	5-30	2-10
Tuff	10-60	5-20
Fresh granite and gneiss	0.01-2	< 0.1
Weathered granite and gneiss	1-15	0.5-5

Sources: Freeze and Cherry, 1979; Todd, 1980; Driscoll, 1986

Residence times in the unsaturated zone depend on its thickness and can vary from almost nothing to tens of years. At greater depths, all the empty spaces are completely filled with water and this is called the saturated zone. If a hole is dug or drilled down into the saturated zone, water will flow from the ground into the hole and settle at the depth below which all the pore spaces are filled with water. This level is the water table and forms the upper surface of the saturated zone, at which the fluid pressure in the pores is exactly atmospheric. Strictly speaking, the term groundwater refers only to the saturated zone below the water table. All water that occurs naturally beneath the Earth's surface, including saturated and unsaturated zones, is called sub-surface water (Fig. 4.6).

A further important way of characterizing aquifers is useful in any discussion of the development and protection of groundwater resources. An unconfined aquifer is one in which the upper limit of the zone of saturation, *i.e.* the water table, is at atmospheric pressure. At any depth below the water table the pressure is greater than atmospheric pressure and at any point above the water table the pressure is less than atmospheric pressure. In a confined aquifer the effective aquifer thickness extends between two impermeable layers. At any point in the aquifer, the water pressure is greater than atmospheric pressure. If a well is drilled through the confining layer into the aquifer, water will rise up into the well until the column of water in the well balances the pressure in the aquifer. An imaginary surface joining the water level in many wells in a confined aquifer is called the potentiometric surface. For a phreatic aquifer, that is the first unconfined aquifer to be formed from the land surface, the potentiometric surface corresponds to the water table found in wells or boreholes. The flow directions of groundwater are perpendicular to the isolines of the potentiometric surface as shown in (Fig. 4.6).

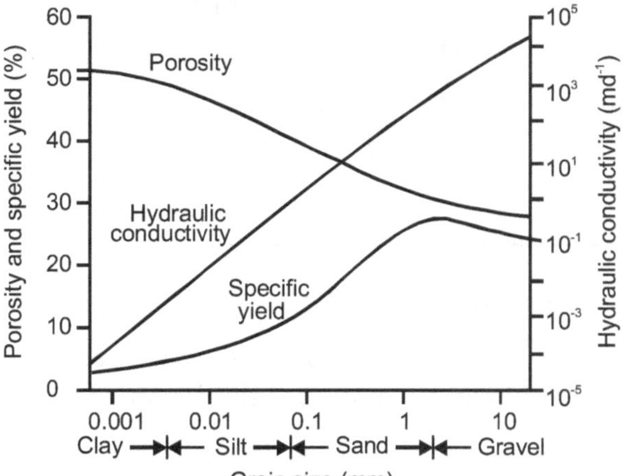

Figure 4.6 Porosity, specific yield and hydraulic conductivity of granular materials (Modified from Davis and De Wiest, 1966)

If the pressure in a confined aquifer is such that the potentiometric surface comes above ground level, then the well will overflow and is said to be artesian as shown in (Fig. 4.2). Clearly, a confined aquifer with thick overlying impermeable clays is likely to be much less vulnerable to pollution than an unconfined aquifer.

The *hydraulic conductivity* (K) of an aquifer is defined as a constant of proportionality relating the specific discharge of a porous medium under a unit hydraulic gradient. Hydraulic conductivity has units of length/time, L/T.

The following table 4.4 shows representative values of hydraulic conductivity for various unconsolidated sedimentary materials (from Domenico and Schwartz, 1990):

Table 4.4: Hydraulic conductivity of unconsolidated sediments

Material	Hydraulic Conductivity (m/sec)
Gravel	$3x10^{-4}$ to $3x10^{-2}$
Coarse sand	$9x10^{-7}$ to $6x10^{-3}$
Medium sand	$9x10^{-7}$ to $5x10^{-4}$
Fine sand	$2x10^{-7}$ to $2x10^{-4}$
Silt, loess	$1x10^{-9}$ to $2x10^{-5}$
Till	$1x10^{-12}$ to $2x10^{-6}$
Clay	$1x10^{-11}$ to $4.7x10^{-9}$
Unweathered marine clay	$8x10^{-13}$ to $2x10^{-9}$
To Convert from m/sec to	Multiply value by
Cm/sec	100
Gal/day/ft^2	$2.12x10^6$
Ft/sec	3.28

Groundwater flow and Darcy's law

The flow of water through an aquifer is governed by Darcy's Law, which states that the rate of flow is directly proportional to the hydraulic gradient:

$$Q = - K\, i\, A$$

Where,

Q is the rate of flow through unit area A under hydraulic gradient.

The hydraulic gradient dh/dl is the difference between the levels of the potentiometric surface at any two points divided by the horizontal distance between them. The parameter K is known as the hydraulic conductivity, and is a measure of the permeability of the material through which the water is flowing. The similarity between Darcy's Law and the other important laws of physics governing the flow of both electricity and heat should be noted. For clean, granular materials, hydraulic conductivity increases with grain size (Fig. 4.6). Typical ranges of hydraulic conductivity for the main geological materials are given in table 4.4 and shown in (Fig 4.6).

Darcy's Law can be written in several forms. By substituting *V* for the term *Q/A* it can take the form:

$$V = K \frac{dh}{dl}$$

in which the term *V* is commonly referred to as the specific discharge. The specific discharge does not represent the actual velocity of groundwater flow through the aquifer material. To determine specific discharge, the volumetric flux *Q* is divided by the full cross-sectional area, which includes both solids and voids. Clearly, flow can only take place through the pore spaces, and the actual velocity of groundwater flow can be calculated if the porosity is known. Most aquifer materials in which inter-granular flow predominates have porosities of 20 to 40 per cent, so the actual groundwater flow velocity is three to five times the specific discharge. However, this average velocity in the direction of groundwater flow does not represent the true velocity of water particles traveling through the pore spaces. These microscopic velocities are generally greater, as the inter-granular flow pathways are irregular and longer than the average, linear, macroscopic pathway. The true microscopic velocities are seldom of interest, and are in any case almost impossible to determine.

Darcy's Law provides a valid description of the flow of groundwater in most naturally occurring hydrogeological conditions. Thus, in general, it holds true for fractured rock aquifers as well as granular materials, although there are particular problems associated with the analysis of flow in fractured media, which are outside the scope of the present discussion. Fissure flow conditions are characterised by low porosity and high permeability, and very high flow velocities may result, especially where a small number of fissures are enlarged by solution (Fig. 4.5e). Extensive development of solution in limestone areas can result in karst terrain, characterised by solution channels, closed depressions, sink holes and caves, into which all traces of surface drainage may disappear. Extreme groundwater velocities of several kilometers per day have been observed in tracer experiments in such areas. These conditions can be very favourable for groundwater abstraction, but aquifers of this type are often highly vulnerable to all types of pollution and difficult to monitor.

Although most of the flow in a fractured rock aquifer is through the fissures, very slow groundwater movement may occur through the interconnected voids of the matrix. An aquifer of this type is characterised by high porosity and low permeability in the matrix with low porosity and high permeability in the fractures. The volume of groundwater stored in the matrix may be 20 to 100 times greater than the volume in the fractures. Many limestones, some sandstone and certain types of volcanic rocks exhibit these characteristics, and such aquifers can be said to have dual porosity and permeability. One of the most intensively used and well-studied aquifers of this type is the chalk of northwest Europe.

The origin of fresh groundwater is normally atmospheric precipitation of some kind, either by direct infiltration of rainfall or indirectly from rivers, lakes or canals. Groundwater is, in turn, the origin of much stream-flow and an important flow component to lakes and oceans and is, therefore, an integral part of the hydrological cycle. Water bodies such as marshes are often transitional between groundwater and surface water. At all places where surface and groundwater meet, the inter-connection between the two must be appreciated in relation to the design of water quality assessment programme.

Within the context of the overall cycle, the groundwater flow system is a useful concept in describing both the physical occurrence and geochemical evolution of groundwater. A groundwater flow system is a three dimensional, closed system containing the flow paths from the point at which water enters an aquifer to the topographically lower point where it leaves. Infiltration of rainfall on high ground occurs in a recharge area in which the hydraulic head decreases with depth and net saturated flow is downwards away from the water table. After moving slowly through the aquifer down the hydraulic gradient, groundwater leave the aquifer by springs, swamps and base flow to streams or to the ocean in a discharge area. In a discharge area, the hydraulic head increases with depth and the net saturated flow is upwards towards the water table. In a recharge area, the water table often lies at considerable depth beneath a thick unsaturated zone. In a discharge area, the water table is usually at, or very near, the ground surface. Rivers, canals, lakes and reservoirs may either discharge to, or recharge from, groundwater and the relationship may change seasonally or over a longer time span.

Water movement in the unsaturated zone cannot be determined by the same methods as for the saturated zone because water will not freely enter void spaces unless the soil-water pressure is greater than atmospheric pressure. Therefore, special methods must be used to measure water movement in the unsaturated zone, and the most common are tensiometers, psychrometers and neutron moisture logging. Groundwater flow in the unsaturated zone is dominantly vertical, and is related to the rate of infiltration, the storage capacity of the vadose zone and the depth to the water table. Downward movement in granular materials is generally a few meters per year or less, and residence times of tens of years are thus possible where the unsaturated zone is thick. In fissured or dual porosity aquifers exposed at the surface, rapid infiltration through the fissures can by-pass the unsaturated zone completely, dramatically reducing residence times and making the aquifer highly vulnerable to pollution.

Recharge, Discharge and Aquifers: When groundwater is replenished, it is recharged. Rainfall, snowmelt and the downward movement of surface water through the soil and upper layers of the unsaturated zone are the main mechanisms for groundwater recharged (Karanth 1989, Lohman 1972 and Todd 1995). Groundwater discharge is the upward movement of groundwater to the surface. This occurs as springs or seepages. A lot of groundwater discharges to rivers. Groundwater usually discharges to rivers, lakes and wetlands because they occupy low areas in the landscape. It is important to note that many wetlands occur where groundwater discharges to the land surface and the water moves from groundwater to wetland before moving on to a lake or stream.

Aquifers are geologic formations that contain groundwater in adequate amounts to be used for commercial, domestic, agricultural or industrial purposes (Lohman 1972 and Todd 1995). Not all groundwater occurs in aquifers, but all aquifers contain groundwater. There are three main categories of aquifers: unconfined, confined and perched.

Springs and Seepages: A spring is ground water that naturally flows from the ground at the surface. Where this flow is diffused it is seepage. Because of the availability of water, many wetlands occur near springs and seepages.

Matrix of aquifer types: It is usual to classify all rock types as igneous, sedimentary or metamorphic, depending on their origins. Igneous rocks are formed by the cooling and solidifying of molten rock or magma, after it has been either forced (intruded) into other rocks below ground or extruded at the ground surface as a volcanic eruption. Sedimentary rocks are formed as a result of the deposition of particles (often resulting from the weathering and erosion of other rocks) usually under water on the seabed, or in rivers, lakes and reservoirs. Metamorphic rocks are formed by the alteration of other rocks under the action of heat and pressure.

Most of the world's important regional aquifers are of sedimentary origin. Igneous and metamorphic rocks are generally far less important as sources of groundwater. As a result of their method of formation, igneous and metamorphic rocks often have such low initial porosity and specific yield (Table 4.3, Fig. 4.6) that they do not form important aquifers. In time, and particularly with elevation to the ground surface and prolonged exposure to weathering, joints may be formed and opened up so that sufficient porosity and permeability is created to allow the igneous and metamorphic rocks to be called aquifers. Locally these weathered rocks can be vital sources of water. In contrast, sediments usually start their geological lives with a high porosity and a permeability, which is closely related to grain size (Figure 4.5). With time, they may be buried and compacted, and cement may form between the grains (Fig. 4.5 d); such consolidated or indurated sedimentary rocks may have low porosity and permeability and only become aquifers by virtue of their joints and fractures.

Table 4.3 represents an attempt to construct a matrix of major aquifer types. The aquifers in the upper part of the table are largely unconsolidated and relatively young in geological terms. They are, therefore, invariably found close to the surface and form aquifers at shallow depths, usually less than 100 to 200 m. Much older sediments, which started life as unconsolidated materials are found at greater depths. There are, for example, ancient glacial deposits many millions of years old but these have been deeply buried and compacted and have lost so much of their original porosity that they can no longer be considered as aquifers. Unconsolidated materials in which inter-granular flow dominates are, therefore, restricted to relatively shallow depths (Table 4.3). Unconsolidated materials are widely distributed and very important as aquifers. They are characterised by generally good natural water quality because of the short to moderate residence times and regular recharge, at least in humid areas, but are often highly vulnerable to anthropogenic influences.

Consolidated formations produce shallow and deep aquifers (Table 4.3). Sedimentary formations originally formed at shallow depths and having high inter-granular porosity may have been deeply buried, compacted, cemented and subsequently brought back to the ground surface by major earth movements. If secondary porosity has developed (Table 4.2), they then form shallow aquifers in which both fissure flow and inter-granular flow may be important (Table 4.3). Such aquifers may be highly productive and have generally good natural groundwater quality if there is regular recharge, but may be especially vulnerable to pollution because of the possibility of very rapid flow in fissures (Fig. 4.6).

Similarly, igneous and metamorphic rocks having low original porosity when they were formed at great depths within the earth's crust can form important (but usually relatively low yielding) shallow aquifers when brought to the earth's surface and exposed to weathering processes. At greater depths, without weathering most of the fissures are closed and insufficient secondary porosity or permeability develops for such rocks to be considered aquifers (Table 4.3).

Igneous volcanic rocks come to the ground surface as molten lava or as ejected fragmentary material. These fragments may be welded by the great heat and gasses associated with volcanic activity and form tuffs. Fine-grained crystalline lavas have almost no primary porosity. Their usefulness as aquifers depends on features such as cooling joints, the blocky nature of some lava types, and the weathering of the top of an individual flow before the next one is extruded over it. Shallow volcanic aquifers have generally good natural water quality for the same reasons as other shallow aquifers, except that in areas of current or recent volcanic or hydrothermal activity, specific constituents, particularly fluoride may present problems.

Deep, consolidated sedimentary formations are characterised by slow groundwater movement, long residence times, ample opportunity for dissolution of minerals and, therefore, often poor natural water quality. As a result of their great depth, such formations are often confined beneath thick sequences of low permeability clays or shale, and are generally less vulnerable to anthropogenic influences.

Chapter 5

Hydrological Properties of Rocks

⇒ This chapter presents the hydrological properties of Igneous, Sedimentary, Metamorphic and Unconsolidated Rocks.

Groundwater exists at many places beneath the surface of the earth. It is normally hidden from view but, on occasion, can be seen flowing from springs and wells or seeping into tunnels. Certain properties of the various geologic materials control the ability of groundwater to enter into, move through, be stored in, or be extracted from the ground. Therefore, an understanding of geology is necessary in order to gain an understanding of groundwater found within a particular area of investigation. A basic part of the study of the geology of an area includes the preparation of a map showing surfacial exposures of each rock type. Interpretation of the map facilitates the study of the underground configuration, characteristics, and physical properties of these materials. This, in turn, enables the evaluation of groundwater storage, movement, and yield, as well as the identification of areas of ground water quality problems, recharge, and extraction.

The fundamental unit of a geologic study is the geologic formation. A formation has been defined as any assemblage of rocks, which have some character in common, whether of origin, age, or composition. In this sense, the term "rock" is defined as any naturally occurring part of the earth's crust, made up of aggregate of minerals and includes hard, dense granite, lava ash, clay, sand, and even loose-uncohesive soil.

The rocks in general have been divided into three major rock groups, *viz.* igneous, sedimentary, and metamorphic. Igneous rocks were formed from the cooling, solidification, and crystallization of molten magma or lava and are of two types, intrusive and extrusive. When formed at depths of several miles, they are spoken of as intrusive. These rocks contain minerals and crystals that can be seen by the unaided eye. When formed on the surface, such as a lava flow, igneous rocks are spoken of as being extrusive. Most crystals in this latter rock type are minute; only a few (phenocrysts) are visible to the unaided eye for example plagioclase in giant phenocryst basalt, a typical extrusive rock found in the Deccan Trap Basalt formations of India. Certain groups of intrusive and extrusive rocks are chemically identical and have inherent characteristics, which are not readily apparent. This is because a single mass of molten magma can produce three strikingly different rocks when solidification takes place under differing conditions. For example, when formed at great depth, a fairly coarse-grained granitic rock will result known as plutonic igneous rock. This same magma, if formed near the ground surface under conditions of lower temperature and pressure, will solidify as a fine-grained rhyolite known as hypabyssal igneous rock. When temperatures and pressures are still lower, the molten mass will chill and obsidian, or volcanic glass will result. Such rocks formed on the earth's surface are called volcanic igneous rocks. This latter rock type contains no crystals.

The second major rock group is the sedimentary rocks. These rocks are formed from the products of erosion of older rock materials. As weathering progresses, erosional products wash down slope and become deposited as gravel, sand, and clay in adjacent valleys or on the bottom of the sea. In time, because of burial and compaction, clay turns to shale, sand to sandstone, and gravel to conglomerate. These are the most common sedimentary rock types; they occur in the Indo-Gangatic alluvial plain (India) and various river basins. Sedimentary rocks also include limestone, a deposit of calcium carbonate, and chert, a deposit of silica.

The third group of rocks is the metamorphic rocks. The term "metamorphic" means changed or altered. This rock type is formed whenever igneous or sedimentary rocks become buried deep within the earth and where pressures and temperatures are extremely high. In such an environment, the minerals, which make up the original rocks, undergo alteration and mixing to form entirely new minerals. Metamorphosed igneous rocks become greenstone and schist, sandstone becomes quartzite, limestone becomes marble, and shale becomes slate.

Metamorphic rocks are found in Archean and Pre-Cambrian rocks of India including Dharwar Super-group, Delhi Group and Archean rocks of Rajasthan, Bihar *etc*.

The geologic formations with reference to hydrogeological properties can be divided into two basic groups, *viz.* water-bearing and non-water-bearing. A water-bearing formation is one that readily absorbs, transmits, and yields usable quantities of ground water to wells. Conversely, a non-water-bearing formation is one that yields only limited quantities of water to wells. In some cases, non-water-bearing formations can yield mineralized, non-potable water. Furthermore, non-water-bearing rocks are all "hard rocks", that is, they are consolidated and massive. Conversely, most of the water-bearing rocks include soft sandstone, clay, alluvial soils, and river gravels.

Nearly all of the materials that make up the water-bearing formations have open spaces containing groundwater. The size of these openings ranges from minute pores in clays to intergranular openings in deposits of sand and gravel. The porosity, or percentage of the total volume of the openings, is not necessarily indicative of the ease with which groundwater moves through the material. If the openings are very small, or if they are not connected, the material has a low permeability. Materials of low permeability, such as clay, transmit very little water. In contrast, materials of high permeability, such as river gravel, yield large amounts of groundwater. A geologic bed or stratum that readily transmits groundwater (*i.e.* has a high permeability) is called an aquifer. In contrast, materials which contain groundwater but cannot transmit extractable quantities (*i.e.*, have low permeability) are called confining beds. Certain strata can act as aquifers in one area and as confining beds in another area because of lateral changes in permeability resulting from changes in the physical characteristics of the materials.

Groundwater exists in two zones beneath the ground surface, such as the upper zone is the zone of aeration and below the water table is the zone of saturation. In the zone of aeration most of the openings in the geologic materials are partly filled with air and partly with water. Wells do not produce groundwater from the zone of aeration because the molecules of water adhere tightly to the various geologic materials. If perched groundwater (Fig. 4.4) occurs in the zone of aeration, it is contained in an isolated saturated zone, which is separated from the main groundwater body by an underlying impermeable stratum the wells in this zone produce a little quantity of groundwater.

In the lower zone, or zone of saturation, all of the interconnected openings in the geologic materials are filled with groundwater; little or no air is present. Groundwater exists in this lower zone under either confined (Fig.4.1) or unconfined (Fig.4.3) conditions. An aquifer containing unconfined groundwater is one that is not overlain by a confining bed. The upper surface of an unconfined body of groundwater is called the water table. The level of water in a well tapping unconfined groundwater is represented by it. Unconfined groundwater moves very slowly in the direction of the downward slope of the water table.

The stratification of aquifers and confining beds is the result of deposition under continually changing environments. Coarse deposits, sand and gravel, are laid down along stream channels. They are coarsest at the apex of alluvial fans and become finer- grained the farther removed they are from the mountains Silts and clays are deposited by slow-moving streams, in flood areas adjacent to active channels, and in lakes, swamps, and bays.

In hard rock area, due to unconfined nature of aquifer system, groundwater targeting, development and management is complex. It is related with climatic, physiographic (surface topographic characteristics, slopes, drainage *etc.*) and aquifer geometry (thickness of weathered horizon *i.e.* depth of basement, fracture zones *etc.*). The groundwater studies remains to be geological bias which may be due to inability of scientists to take into accounts of other parameters in hydrological prospective and also may be due to tendency to find quick solution as other natural resources mapping are done *i.e.* landuse mapping using remotely sensed data. Due to this tendency, trend has emerged to depict the groundwater potential on the basis of geomorphology using remotely sensed data, which has been termed as hydrogeomorphology.

The hydrological properties of different rock groups are described below:

Igneous Rocks

Igneous rocks having low original porosity when they were formed at great depths within the earth's crust can form important (but usually relatively low yielding) shallow aquifers when brought to the earth's surface and exposed to weathering processes. At greater depths, without weathering most of the fissures are closed and insufficient secondary porosity or permeability develops for such rocks to be considered as aquifers.

On the basis of mode of occurrence igneous rocks are classified into three types such as volcanic, hypabyssal and plutonic igneous rocks as described earlier (para 3 this chapter). Based on silica percentage the igneous rocks are classified into three types *viz. acidic*, when silica percentage is >66% (*e.g.* Rhyolite, Granite and Pegmatite); *intermediate*, when silica percentage is between 66% and 54% (Syenite, Diorite and Aplite) and *basic*, when silica percentage is <54% (Basalt, Gabbro and Dolerite). Depending on feldspar content the igneous rocks are further classified into alkali series *e.g.* granite (when K or Na feldspars like orthoclase and albite are dominant in the rock) and calc-alkali series *e.g.* basalt (when Ca plagioclase feldspar is dominant in the rock).

Igneous volcanic rocks come to the ground surface as molten lava or as ejected fragmentary material. These fragments may be welded by the great heat and gasses associated with volcanic activity and form tuffs. Fine-grained crystalline lavas have almost no primary porosity. Their usefulness as aquifers depends on features such as cooling joints, the blocky nature of some lava types, and the weathering of the top of an individual flow before the next one is extruded over it. Shallow volcanic aquifers have generally good natural water quality for the same reasons as other shallow aquifers, except that in areas of current or recent volcanic or hydrothermal activity, specific constituents, particularly fluoride may present problems.

In igneous rocks, the restricted opportunity for reactions to take place is accentuated by the fact that groundwater storage and flow is predominantly in fissures, giving short residence times and low contact surface area. Groundwater in igneous rocks is, therefore, often exceptionally lightly mineralized, although characterised by high silica contents (Hem, 1989).

The plutonic (intrusive) igneous rocks are dense in texture with all component minerals very closely crystallized so that there are no pore spaces or openings. These rocks are non-porous, impermeable and without any scope of holding groundwater. The only possibility of such rocks to hold the groundwater is the development of joints and fractures in the rocks due to folding, faulting and other forces. On the other hand, extrusive igneous (volcanic) rocks exhibit great variation in their water bearing character. The composition of these rocks and their mode of formation from lava as well as the nature of original topography are factors, which define the water bearing capacity of these rocks.

Plutonic igneous rocks like granites are acidic, coarse grained and crystalline hard rocks do not contain the primary porosity. The porosity in the rock is only the secondary porosity developed due to the joints and fractures. Hence the groundwater in these rocks is restricted mostly to the zones of secondary porosity developed due to fractures, joints and weathering profile.

Volcanic acid igneous rocks such as Rhyolites etc may or may not be holding water because acidic lavas are comparatively viscous and fragmentary at the time of eruption. Though interstices and fracture system are common in such lavas, but they are either filled with ash and other earth materials like clay particles thereby reducing the porosity and permeability. Volcanic basic igneous rocks such as basalt are characterised by a great gas content and higher mobility of lava at the time of eruption. During cooling of the lava these rocks contain cavities (due to escape of gases during cooling) and cracks and joints (due to contraction on cooling). These cavities and cracks may impart the porosity and permeability to this group of igneous rocks.

In basalt the groundwater occurs under phreatic conditions in the exposed lave flows and under semi-confined conditions in the flows at deeper level. Lithological constraints dictate that groundwater is present in the pore spaces of the vesicular basalt and in the jointed and fractured portions of massive parts of the flows. The primary porosity in the basalts is associated with the vesicles, which are the pore spaces developed due to the escape of volatile and gases when the lava erupts on the surface as a lava flow. This primary porosity in the basalt is naturally limited and related to the quantum of gasses/volatile in the eruptive phase, which resulted in the basalt flow. The groundwater in these rocks therefore is restricted mostly to the zones of secondary porosity developed due to fractures, joints and weathering.

From the hydrogeological point of view, the frequency and extent of jointing, fracturing and the flow contacts and weathering along them are the most significant parameters imparting permeability and porosity for forming suitable groundwater reservoirs in the basaltic terrain. The vesicular zones occurring in the upper parts of flows or units, though porous, are not permeable, as the vesicles are not interconnected. Secondly, the vesicles are generally filled with amygdules, green earth, glassy material *etc.* The red bole layer, flow breccia with secondary mineral development and the massive parts of the flow, with non-interconnected joints, are impervious (Thorat and Sable, 1990). The secondary porosity (Joints and fractures) generally reduces with depth and hence the near surface (unconfined) aquifer system rarely, extends below 30 m depth (Agashe, 1990).

Hydrogeological characters of basalt flows have indicated that the two distinct types of lava flows have distinct qualities as for as their porosity and permeability are concerned. These are amygdaloidal basalt flow and compact basalt flow as described below:

In amygdaloidal basalt (Pahoehoe type) flow the original gas cavities are filled up with secondary minerals obliterating original vesicular nature. Due to presence of amygdules, fresh amygdaloidal basalt is free from joints and occurs as homogeneous, watertight mass (Patil *et al,* 1999). The vesicular amygdaloidal basalt unit is more susceptible to weathering and exhibits deep weathering profile (Fig 5.1). It is characterised by the formation of sheet joints. Sheet joints are also developed at the contact between the vesicular amygdaloidal and the compact basalt that is with in the contact zone (Fig. 5.1). Such weathered amygdaloidal basalt contains groundwater. (Deolankar, 1980; Lawrence, 1985 and Kulkarni and Deolankar, 1995 and 1997). However, quantity of groundwater depends upon the thickness of weathered zone.

The compact basalt (aa type) flows can be demarcated into two parts according to their hydrogeological characters. The top portion of this flow is vesicular, unjointed and watertight in fresh condition (Fig. 5.2), but produces sheet jointing due to weathering. The middle and lower parts of the compact basalt flow are jointed (Fig. 5.2).

Fig. 5.1: Vertical section in the well at Raipur near Pingalgad nala stream showing weathered amygdaloidal basalt (WAB), compact basalt (CB) and sheet joints (SJ) at the contact of the two flows.

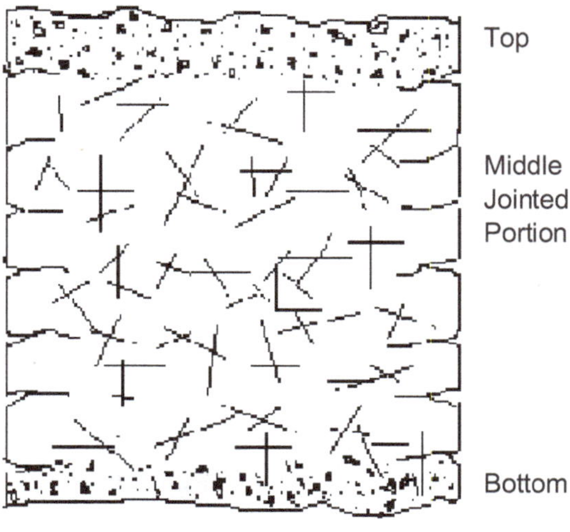

Fig. 5.2: Typical Sketch of Compact Basalt Flow.

Fig. 5.3 Dug well in compact basalt.

Illustrates the dug well in jointed compact basalt flow. These joints in the compact basalt are closely spaced linear along preferred orientations and the continuity of these joints is traceable over considerable distances. Such fracture zones, which have a discrete distribution in space and orientation, have been reported and described by various workers (Powar, 1981; Peshwa *et al,* 1987, Kale and Kulkarni, 1992 and Kulkarni, 1992). Such fracture zones transect several Deccan basaltic units and constitute recharge conduits (lateral as well as down ward transmission of water) for deeper Deccan basaltic aquifers (Kulkarni *et al.,* 1994).

However, quantity of percolation of water depends upon joint spacing and pattern of jointing (Fig. 5.4). Water can percolate through closely spaced joints faster as compared to broadly spaced and non-interconnected joints.

Sedimentary Rocks

Sedimentary rocks form by the lithification of mineral and rock pieces, through precipitation from water, or through biochemical production. Although sedimentary rocks comprise only 5% of the crust, they represent nearly 75% of surface outcrop: they are the rocks we usually see. In addition, they are the ultimate source and storehouse of fossil fuels and many valuable natural resources, such as iron ores and fertilizers.

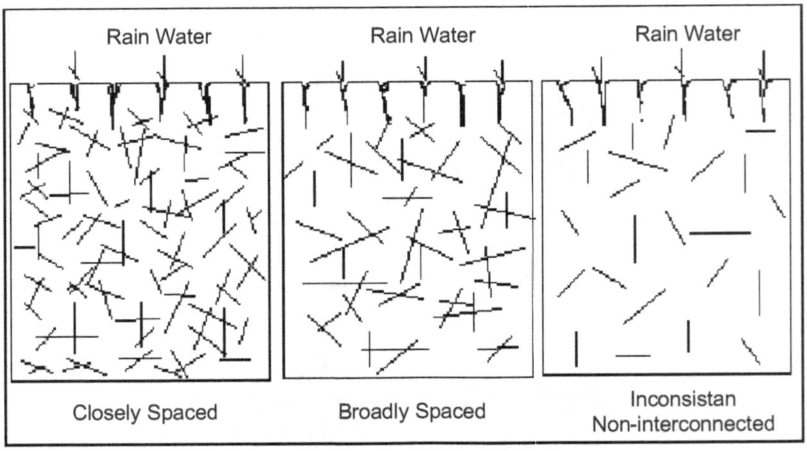

Fig. 5.4: Schematic diagram showing jointing patterns in Basalt.

In broad outline, the production of sedimentary rock follows the following steps:

- Weathering fractures and dissolves rock, creating sediment and dissolved components.
- Erosion carries away the products of weathering.
- Wind, water and glaciers carry them to where they are deposited.
- Deposition accumulates sediments from large source regions.
- Burial within the sediment pile compacts the sediments.
- Diagenesis transforms the sediments into sedimentary rock.

Sediment becomes sedimentary rock through the process of *diagenesis* or *lithification*. This begins with the compaction of sediments with continued deposition and burial. Pore space in unconsolidated sediments is often as high as 60 or 70%. Compaction reduces this to 10 or 20%. Lithification occurs with the precipitation of dissolved mineral content forming cement holding sediment grains together. This is frequently calcium carbonate in the form of Aragonite or Calcite, but silica is also found. With increasing burial depth comes increasing pressure and temperature, driving diagenetic changes that occur as crystals adopt more stable structures. Aragonite converts to Calcite, clays experience a suite of transformations, and some grains will recrystallize at contact points where pressures are highest, further lithifying the rock. High pressures and temperatures also cause the breakdown of organic materials, starting the series of reactions that lead to coal or to oil and gas.

Sedimentary environments

Sediments are deposited in a variety of places. Each is characterized by a combination of the kind and amount of water, the local topography, and the biological activity. These in turn are related to plate tectonic setting and climate. Although sediment can be deposited anywhere, large accumulations of sediment are not, being found in only a handful of environments. In discussing these we will find it helpful to break them down into continental, shoreline and marine settings.

Continental: The primary continental sedimentary environment is the *alluvial* environment, consisting of streams and flood plains. Rivers occur everywhere, making this a common environment. All climates have alluvial deposits, though the rivers may be ephemeral. They are characterized by efficient downstream sorting and deposition of fine load in floodplains, which are also rich in organic matter. *Lacustrine* (lake) environments are characterized by weak currents and deposition of fine sediments (silts and clays), abundant organic material, occasional chemical deposits (both carbonates and silica), and, in arid climates, evaporites (salts). *Swamps* are stagnant, usually fed by weak currents and/or rain. They are rich in organic material, sometimes producing peat. *Deserts* are obviously arid, making wind the primary transportation agent. Fluvial (stream related) processes are intermittent or absent. So too is biologic input. Last are *glacial* environments. They are cold and dominated by ice dynamics. The resulting deposits demonstrate the little sorting or shaping of sediment and no stratification (no obvious layering of sediments). Like alluvial environments, glacial deposits range from fines to very coarse sediments.

Shoreline: *Deltaic* environments occur where rivers dump into lakes or oceans. Here river currents, waves and tides interact, producing complex deposition and sorting patterns. Because of the high dissolved and suspended load, there is usually strong biologic and organic input. Sediment input is very high. *Tidal flats* are mixed subaerial and tidal (alternately on land/off land) and dominated by weak tidal currents.

Dominant sediments are sands and mud. *Beach* environments are high-energy, driven by waves and tides. Mud is kept in suspension, leaving sands and gravels. Shaping and mechanical breaking of grains is efficient, resulting in mature sediments.

Marine: The *continental shelf* is the gently sloping surface that extends from the shoreline to the shelf break at a depth of about 150 m. Waves and tides dominate, but because of the water depth, the currents are weak. Sand and mud are deposited and are chiefly terrigenous in origin (derived from continents). Coral reefs and carbonate mud also accumulate here. The *continental margin* outward from the shelf is subject to ocean and turbidity currents: flows of dense water and sediment mixtures driven by gravity. These bring mud and sand down from the shelf. Biologic activity contributes shells and tests that rain out of the water column. *Deep-sea* sediments are primarily mud and biogenic sediments. The latter are referred to as *pelagic*, referring to the shells of foraminifers, diatoms and radiolarians, and also consisting of carbonate and silica. Ash from eruptions and dust from high winds also finds their way here. Currents are weak, allowing these fine particles to settle out.

Within a sedimentary environment, we refer to the total of identifiable characteristics as a *facies*. This includes grain size, the degree of sorting, grain shape, mineralogic breakdown and the structures within the deposits. Facies distinguish between different sedimentary environments that co-exist. For instance, at a shoreline you could move landward from a beach, to a backwater lagoon to an alluvial setting, each a facies, each imprinting identifiable marks on the sediments deposited there. Facies associations like these are diagnostic of changing environments, such as rising or falling sea level.

Hydrological conditions of some common sedimentary rocks are described below:

Sandstones: The sandstones show great variation in their water yielding capacity, which is chiefly controlled by their texture and nature of cementing material. The coarse grained sandstones with rather imperfect cement are excellent aquifers, whereas fine grained varieties of sandstone with thoroughly compacted or cemented may prove useless as water yielding rocks. In permeability, sandstones vary from semi-permeable to highly permeable, with porosity ranging from 5 – 30 percent, specific yield from 5 – 15 per cent and hydraulic conductivity between 1 – 100 m/day.

Pure siliceous sands or sandstones without soluble cement also contain groundwater with very low total dissolved solids (Matthess, 1982). In such aquifers, the dissolved constituents, which are present, come mainly from other sources, such as rainfall and dry deposition, especially sodium, chloride and sulphate, which in coastal regions, may exceed calcium, magnesium and bicarbonate. Sulphate may also be produced by the oxidation of metallic Sulphides, which are present in small amounts in many rock types. The presence of soluble cement may produce increased concentrations of the major ions.

Carbonate rocks:

The best known productive aquifers of the world are carbonate deposits such as limestones and dolomites. In these rocks the availability of groundwater is controlled by the secondary fractures, solution channels, cavities, caverns and similar openings developed in them rather than by the primary porosity. The porosity of these rocks varies from 1 – 20 per cent and hydraulic conductivity from 1 – 100 m/day. The hydraulic conductivity in highly karsted (cavernous) limestone is may be as high as 100 m/day or even more, so it is classed as the extremely permeable rock. Groundwater in carbonate rocks has pH values above 7 and mineral contents usually dominated by bicarbonate and calcium.

Shale:

This group includes highly compacted and very fine-grained clay, mudstone, marl and shale rocks. They do not possess the primary porosity. These rocks act as the impervious layers. The weathering and secondary porosity may induce some amount of water holding capacity. The porosity of these rocks vary from 0 – 10 per cent and permeability less than 0.1.

Metamorphic Rocks:

Metamorphism describes the set of solid state processes, which transform one type of rock into another. Any type of rock, whether igneous, metamorphic, or sedimentary, may be metamorphosed. Metamorphic changes occur mainly in mineral structure and texture; changes in chemical composition may also take place. Metamorphism typically occurs at the high temperatures and pressures, which are present deep within the earth's crust. The process is defined to occur in the solid state and melting of the rock may therefore not occur. Any process during which complete melting of the rock does occur is considered to be of igneous rather than metamorphic nature.

Metamorphic processes may be divided into two categories according to the area of their extent. Regional metamorphism consists of geologic processes, which act over wide areas and cause metamorphic changes in vast expanses of rock. In contrast, local metamorphism affects only relatively small areas. Local metamorphism occurs when an igneous body intruded into and heats the surrounding rocks. For this reason local metamorphism is also called contact metamorphism.

Different types of metamorphism may alternately be categorized according to the agents, which are active during each process. Stress, pressure, and shock are causative agents of dynamic metamorphism, while thermal metamorphism is due merely to the influx of heat from a nearby igneous intrusion. In contrast, hot, volatile fluids play a significant role in hydrothermal metamorphism. Hydrothermal metamorphism takes place when volatile solutions percolate into and react with a host rock. Both the heat of the intrusive igneous body and the reactive fluids serve to catalyze metamorphic reactions in the host rock. The intrusive magma and the associated volatile fluids may introduce previously absent elements into the reaction process; the chemical composition of the host rock may therefore be extensively altered during hydrothermal metamorphism.

Generally speaking, the crystalline and compact metamorphic rocks like quartzites, marbles and gneisses are non-porous and impermeable holding no prospects for groundwater reservoir. The only exception is when they are profusely fractured and fissured, in which case their behaviour is similar to igneous rocks of plutonic nature. Groundwater will be found in the fractured zones without lateral and vertical continuity.

The highly fractured and foliated metamorphic rocks such as schists, phyllites slates and often some may have been proved to be good aquifers, in a broader sense. As these rocks are layered, folded and faulted, they may have the better chances as aquifers.

Unconsolidated Rocks

Unconsolidated rocks includes various loose fragmentary material before sedimentation or consolidation *i.e.* gravel, sands, tills, *etc.*

Gravels: These are the loose fragments of coarser dimension of gravel and pebble grade. They are the best formations yielding the groundwater. But they also show great variation in the water bearing properties depending on degree of sorting of grains, packing and cementing of the grains if of older age. Generally, the gravels are coarser and well sorted material, which has a very high effective porosity and permeability. Hydraulic conductivity (K) of pebble gravels free from interstitial packing may range between 100 – 1000 m/day. The porosity of these gravels range between 24 – 40 per cent and specific yield between 15 – 30 per cent.

Sands: Sand beds or loose sandy layers are rated as excellent aquifers next to gravels as water yielding materials. The sand layers are characterised with a porosity of 25 – 40 per cent and specific yield of 15 – 25 per cent. The hydraulic conductivity is also quite high ranging from 10 to 100 m/day. In some cases these aquifers proved to be better than gravels because of their extensive continuity. The beach sands are examples of best water-bearing formations in many regions. Sands after compaction and cementation form the sandstone, which is a different aquifer.

Glacial deposits (Tills): Outwash plains deposited by glacial melt waters and tills and tillites developed during last ice age cover a good part of the globe in the northern hemisphere. When these accumulations are made up of sediments of uniform size and rounded nature, they are the good aquifers. When sufficiently thick and extensive these may be very reliable groundwater reservoirs. The specific giving yield varies from 5 – 10 thousand m/day.

Chapter 6

Structural Hydrogeomorphology

⇒ This chapter includes the hydrogeomorphological characters of uniclinal structures and hydrogeomorphic expressions of folds, faults, joints and lineaments.

Application of the principal of geomorphology provides information, which will be of value in predicating the geometry of aquifers. On weathering and erosion, many geological formations develop landforms that are distinctive in respect of slope continuity of outcrops and symmetry of valley flanks. The surface topographic features of bedrocks can some times be extrapolated to reasonable depths to predict the thickness of alluvium or aeolian sands occurring as valley fill deposits, by treating slope profile as mathematical curves for which equations similar to regression equations can be found (Doornkamp and King, 1971).

The influence of tectonic movements and resultant structural features on landform is of paramount importance for hydrogeomorphologists. The control of structure on landforms and hydrogeological processes are being discussed here. The structural geology concerns with the features such as uniclinal or inclined beds, folds, faults and joints. Of these the structural features like faults, fractures, joints *etc* appear as linear feature.

Linear features on the surface of the earth have attracted the attention of geologists for over one hundred years. This interest has grown most rapidly since the introduction of aerial photographs into geological studies. Geologists have recently proven that various structural features perceived in remotely sensed images are reliable indicators of geologic resources (Caran *et al,* 1982). These structures have been used in many applications: petroleum and mineral exploration (Blanchet, 1957); nuclear energy facility siting (Seay, 1979); geothermal assessments (Woodruff *et al,* 1982); and water resource investigations (Lattman and Parizek, 1964). Lattman and Parizek (1964) established a relationship between the occurrence of groundwater and fracture traces for carbonate aquifers, and in particular zones of localized weathering and increased permeability and porosity underlying these structures.

The origin of springs coincides with the existence of a structural or stratigraphic discontinuity (Early, 2001). These discontinuities allow for an increase in aquifer permeability and in turn, groundwater flow rates.

Hydrogeomorphology of Uniclinal Structures

Uniclinal structures are those, which are represented by inclined rock strata at uniform dip angle caused by general regional tilt. These structures are formed in two main ways either by the uplift of a sequence of off-lapping coastal plain sediments or as part of one limb of a large dome or fold (Chorley *et al,* 1985). According to Small (1970), 'uniclinal structures are those in which a general regional tilt has been given by gentle earth movement to the constituent rocks'. Such structures involve both resistant and soft rocks and some times there are alternate bands of soft and resistant rocks and hence these are subjected to differential erosion wherein resistant rocks are less eroded than soft rocks.

The differential erosion of dipping strata of varying resistance produces trellis drainage pattern and a few typical topographic features such as scarp and vale topography, cuesta and hogback ridges *etc.*

Ollier (1981) illustrated the development of structurally controlled streams on dipping strata (Fig 6.1). According to him tributaries join the major consequent or strike streams almost at right angle. The streams flowing the down the dip slopes are called dip streams while the streams flowing in anti-dip direction are called anti-dip streams (Fig 6.1). It may be pointed out that dip streams drains on resistant rocks while anti-dip streams are developed on less resistant (soft) rocks. The relative length of dip and anti-dip streams depend on the angle of the dipping strata. If the dip angles are relatively gentle, the slope lengths becomes longer and hence streams draining on dip slope are longer in lengths than the tributaries draining opposite direction — *i.e.* anti-dip streams. The drainage density on dip slope and anti-dip slope is also variable as anti-dip side of the ridges is characterised by closely spaced streams of relatively shorter lengths have higher drainage density, while relatively low drainage density due to relatively longer length but widely spaced streams on dip slope.

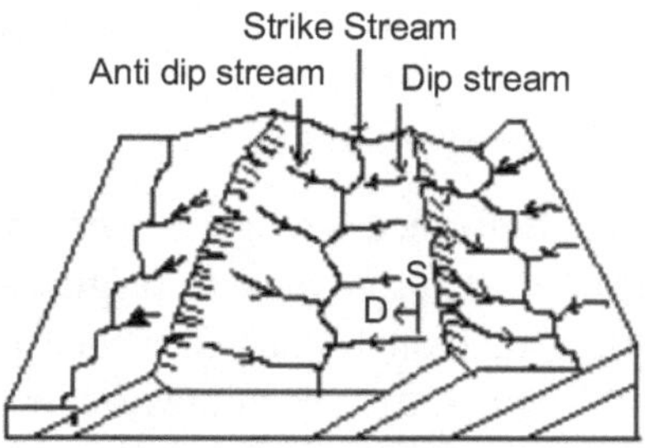

Fig. 6.1: Development of structurally controlled streams on dipping strata (After Ollier, 1981).

Thus it is evident that the structural control of tilted strata imposes a powerful asymmetry on drainage networks. Escarpment streams are steep, short and have high gradients. Dip slope streams are likely to have gentler gradient, more tributaries, large watersheds and more sustained flow (Bloom 1978). Cuesta is the most significant landform resulting from continued erosion of uniclinal or homoclinal sedimentary structures alternated by resistant and soft rock beds.

Striking examples of distinctive geomorphology of rock formations may be found in the development of symmetrical U- or V- shaped valleys with steep slope in steeply dipping quartzite-schist sequence of Delhi system (Fig. 6.2) and asymmetrical valleys in the gently dipping shale – sandstone sequence of Vindhyan super group (Fig. 6.3). When these formations form bedrocks under alluvial cover, the extent of the alluvium, saturated thickness and aquifer boundary conditions can sometimes be predicted. Maximum thickness of aquifer is normally found in the middle valleys of Delhi rocks and near the flank with steeper slopes in the valleys of Vindhyan rocks. Successful tube wells of 20 to 50 m depth, with discharge of 100 m^3/h or more can be constructed at distances of 50 to 100 m quartzite ridges with steep slope usually ranging from 30^0 to 45^0.

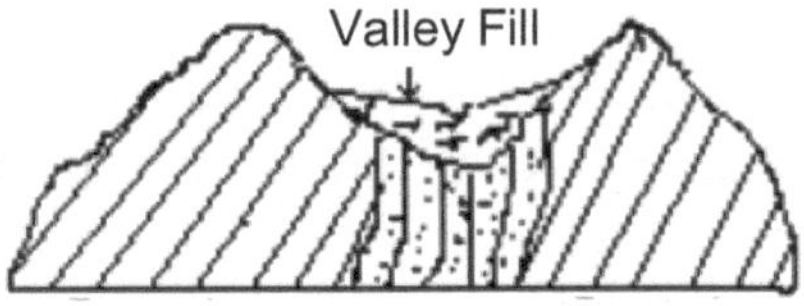

Fig. 6.2: V- shaped valley in steeply dipping strata.

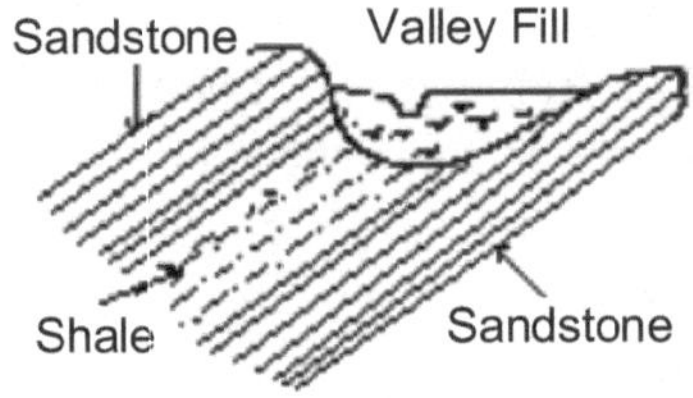

Fig. 6.3: Asymmetrical valley in the gently dipping strata.

Hydrogeomorphic Expression of Folds

Fold, is the bent or deformed arrangement of stratified rocks. These rocks may be of sedimentary or volcanic origin. Although stratified rocks are normally deposited on the earth's surface in horizontal layers, they are often found inclined or curved upward or downward. Arches or up-folds in stratified rock are called anticlines while depressions or down-folds are called synclines.

The nature of the original fold can generally be determined from the arrangement of the outcrops or exposed portions; thus, two outcrops dipping toward each other mark a syncline and two outcrops dipping away from each other an anticline. Folds on a grand scale, extending, for example, most of the length of a continent, are known as geosynclines and geanticlines. The immediate cause of folding is generally conceded to be the horizontal compression of the earth's surface, anticlines being squeezed up by this compression and synclines formed between anticlines. The problem of the ultimate cause of fold formation is similar to that of fault formation, both being earth movements involved in mountain building and plate tectonics. Porous and permeable rocks of anticlines often contain oil and gas reservoirs. Organic remains of late Paleozoic tree fern swamps were converted to anthracite coal during the folding of the Appalachian Mountains.

Fold geomorphology includes the development of drainage pattern and topographic features due to denudational processes on folded structures. One of the resultant features of prolonged erosion of folded structure is the development of inversion of relief *i.e.* inverted relief characterised by anticlinal valleys and synclinal ridges (Savindra Singh, 1998). Inversion of relief in folded structure is an important but unique phenomenon, which causes reverse sequence of topographic features.

Inversion of relief occurs in the folded structure having symmetrical folds having alternate sequence of anticlines and synclines and simple formation (Fig 6.4). Savindra Singh (1998) very well illustrated the stages of inversion of relief from anticlinal and synclinal folds (Fig 6.4 1-5). With the initiation of fluvial erosion under the process of cycle of erosion after the folding of sedimentary rocks longitudinal master consequent steams parallel to strike and tributary consequent streams following slope direction are originated in the synclines and dip slopes of the anticlines respectively. The master consequent stream flows in the syncline from higher slope towards lower gradient. The streams originate on the flanks of the anticlines (dip slope) join the master stream as tributary. These tributaries are called transverse consequent streams. With the passage of time crest of the anticlines breached and subsequent streams develop along the axes of anticlines.

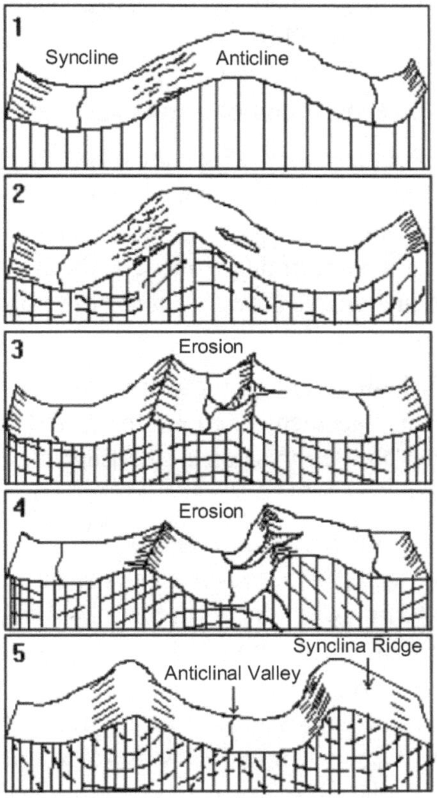

Fig. 6.4: Stages of inversion of relief (After Savindra Singh, 1998).

These subsequent streams continues to deepen valleys due to maximum vertical erosion of anticlinal crest because of maximum tension on crests with the result of synclinal master consequent streams are eliminated and anticlinal streams becomes master streams. This process results in the formation of valleys in place of anticlines and ridges in the place of synclines. Thus, the previous topographic features [Fig. 6.4 (1 and 2)] of original anticline and synclines are reversed by the formation of synclinal ridges (in place of original synclines) and the anticlinal valley (in place of original synclines, Fig. 6.4 (5)) due to prolonged denudation and the process of inversion of relief is completed.

Structural influence on groundwater flow (White, 1999) suggests that conduits tend to be oriented along structural strike in regions of strongly folded rocks. The nature and lateral and vertical extent of aquifers are controlled by the lithology, stratigraphy and structure of the rock formation. The structural features like fold are the manifestation of repetition and reversal of beds.

Hydrogeomorphic Expression of Faults

A fault is a fracture along which there is a displacement caused due to tensional forces. The rocks can be displaced along a plane known as fault plane. The faulting involves the differential movement of strata on either side of fault plane or fault zone. The differential movement in the fault may be upward, downward, horizontal, oblique or rotatory (Small 1970).

The study of fault geomorphology involves 3 aspects of faulting *e.g.* (a) types of displacement of rock blocks and thus the resultant fault type; (b) tectonic expression of faulting and (c) geomorphic expression of faulting.

Based on various types of movement — *i.e.* displacement different types of faults are created in the crustal block such as normal and reverse fault, dip or strike slip fault, transform fault, step fault, *etc* (Fig. 6.5). The tectonic expression of fault includes various types of fault scarps such as active fault scarp, residual fault scarp and composite fault scarps (Fig.6.6). Geomorphic expressions resulting from differential erosion of fault scarps and upthrown and downthrown fault blocks include different types of fault line scarps *viz.* (a) normal or consequent fault-line scarp; (b) reversed or oblique or opposite fault-line scarp; (c) resequent fault-line scarp; (d) subdued fault-line scarp; (e) exhumed fault-line scarp; (f) exaggerated fault-line scarp, *etc.*

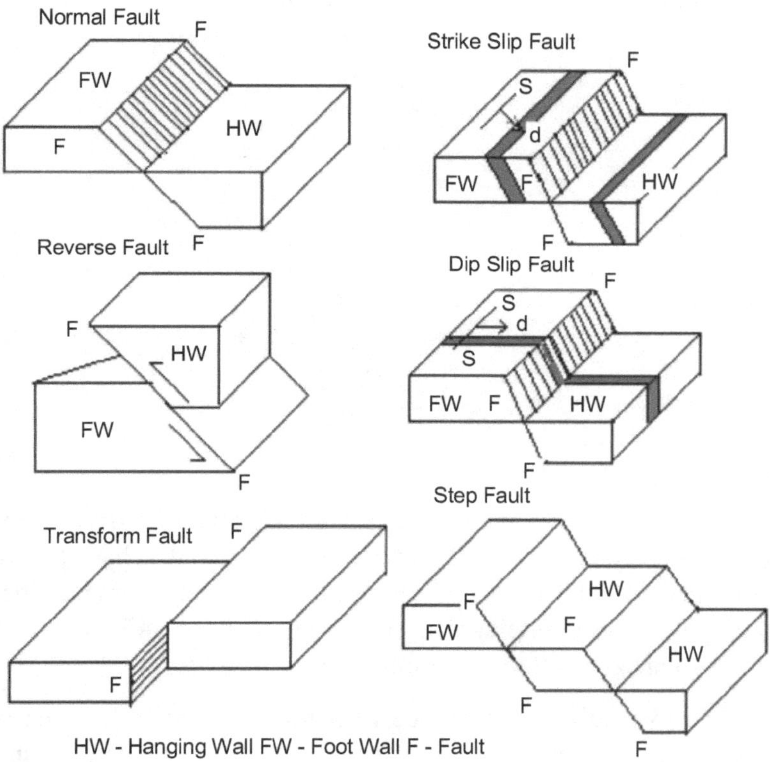

Fig. 6.5: Types of Faults.

Fault displacement is predominantly oblique-slip. A strike-slip component of displacement can be inferred from geometries of fault systems and associated secondary structures. Fault plane striations can also be observed infrequently. The structure of the shear zone associated with a fault also demonstrates the common occurrence of a strike-slip displacement component.

Enhanced development of small faults is observed adjacent to the shear zone of a large fault as compared to a location more distant from the fault. The orientation and density of small faults in this zone reflects the local strain field, which can differ from areas further from the fault. The nature of this small-fault system can vary along a fault. For example, large faults that evolve by linkage of a set of en echelon, overlapping segments locally have greater densities of small faults in the displacement transfer zones between overlapping segments as compared to a position adjacent to a non-overlapping segment. This case exemplifies the western bounding fault of the graben.

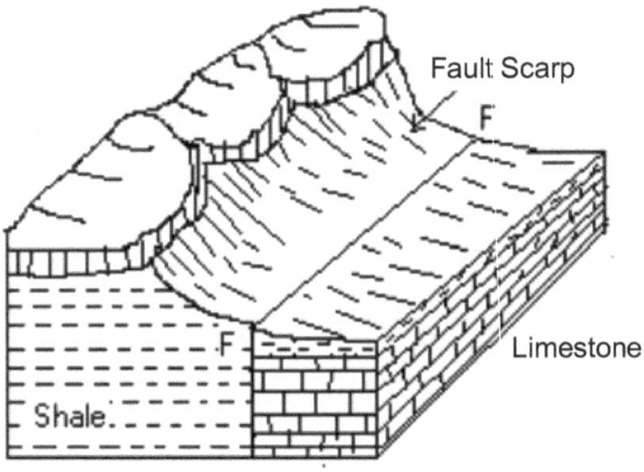

The net thickness of fault gouge associated with the shear zone of a large fault depends upon the width of the zone and the characteristics of the small faults that comprise the zone. Observed shear zones have small fault spacing ranging from 1 to 3 cm, with the individual faults having gouge zones from 0.1-1 cm thick.

The hydraulic conductance of a fault should reflect the net thickness and permeability of fault gouge associated with both the shear zones of the fault and the neighboring zone of associated small faults. Consequently, one would like to know the net thickness of gouge and its spatial variation along a fault.

According to (White, 1999) active faults act as paths of high permeability and concentrated groundwater flow and stable faults have little influence on flow paths due to secondary mineral deposition. These generalizations will aid in establishing relationships between springs and linear features.

Downey (1984) evaluated the geologic conditions and rock properties that produce a good seal when dealing with hydrocarbons, and some of these characteristics can be applied to evaluating the sealing properties of faults in relation to the trapping of water to form perched aquifers. Faults trap not so much by the properties of the fault itself but rather by placing a porous, permeable stratum adjacent to an impermeable stratum across the fault. However, an important instance where the fault itself may play a vital role in transporting or inhibiting water is where there is an overall tensional regional stress field that causes the fault to be much more likely to act as a conduit for flow. The minimum pressure required to displace connate water from void spaces of the fracture, thereby allowing leakage, determines the sealing quality of a fault. The capillary pressure or sealing capacity of a rock is a function of the water-air interfacial tension, wettability, and size of the largest opening. The water-air interfacial tension is a constant, so the variability of the sealing capacity is dependent upon the remaining two properties, with the sealing capacity increasing as the size of the largest opening and the wettability decrease. In order for a seal to exist the total head pressure must be matched or exceeded by the capillary forces. In addition, the average sealing capacity is not the most important criterion to be met; rather, the least sealing capacity must match or exceed the total head pressure, because even a relatively small leak can allow a large amount of water to escape over a significant time period. Some of the conditions that affect the sealing property of a fault are orientation of the strata with respect to the fault, the regional stress field (as mentioned above), the offsetting of geo-pressures, and the effects of the fault movement on the geologic units. For example, differences in geo-pressures across a fault may enhance or diminish the sealing value of the fault. Within faulted blocks, pressure gradients usually run parallel to lithostratigraphic boundaries, and movement along the fault may offset the pressure gradients across a fault. Thus when high-pressured, impermeable strata are juxtaposed across the fault from and in the down-dip direction of permeable, lower pressured strata, the sealing potential at the fault is enhanced. Another example is the case when under certain circumstances thick, under-compacted shales are interspersed between permeable layers. Then, movement along the fault may produce clay "smears" on the ends of permeable layers that inhibit movement of water across or along the fault (Downey, 1984).

When a permeable stratum overlays an impermeable stratum, the rotation of faulted blocks would cause any downward infiltrating water to be channeled in the down-dip direction. If the flow in the down-dip direction is inhibited (*e.g.,* the juxtaposition of the impermeable layer against the permeable layer across a fault), a structural trap is produced that would permit the accumulation of infiltrating water and allow a perched aquifer to form. Perching will also depend on whether the fault/fracture zone acts as a barrier to fluid flow or a conduit. Preliminary numerical analyses of the role of fault zones have been presented by Wittwer *et al.* (1993), Bagtzoglou *et al.* (1994), and Ofoegbu *et al.* (1999, 2001) who indicated that, depending on the properties of the surrounding material, fault zones can interchangeably act as both barriers and conduits to flow. Similar behavior was observed by Berg and Haveman-Avery (1995) in Tertiary growth faults in the Texas Gulf coast. These faults were found to act as seals in the sheared zone and leak along the fault surface. Mineralogical changes within or between strata may also cause perched water due to a change in permeability. Such mineralogical changes within or between strata may also cause a change in permeability due to changes in volume during mineral formation. One example is the formation of relatively impermeable layers in response to alteration of primary phases, such as feldspar to clays like kaolinite and smectite. Trends in mineralogy of this type point to areas of focused flow, and help to delineate palaeo-hydrology. High concentrations of clays such as smectite may also serve to create low permeability zones that are favorable to the formation of perched water zones. Perched water bodies would be more likely to form or expand under future, wetter, climatic conditions that cause increased recharge. Perched zones may also form where alluvial deposits underlie intermittent stream channels. When ephemeral stream flow saturates the alluvial material, the alluvium may remain saturated long after the stream channel is dry at the surface. Even though these perched zones would be located near the surface, the water could eventually drain through fractures to the water table below. More importantly, it is possible that water from these zones could find its way into the host rock by vertical and lateral flow along contacts at, or near locations where the waste containment horizon unit is exposed. Perched zones in alluvial material may be difficult to identify because they are probably infrequent and of short duration. However, like intermittent streams, they have the potential to form, drain, and form again.

Thus it can be noticed that faults can affect groundwater in one of three general ways: (a) enhance flow rates, (b) inhibit flow, or (c) act in a neutral manner. Since this first part of a two-part paper is of a methodological nature, faults are considered to be neutral. The only effects of the relative positions of the layers of different, and perhaps sharply contrasted, hydrologic properties would affect the flow.

Hydrogeomorphic Expression of Joints

Geological survey, drilling and mining often show that the rocks, especially crystalline, volcanic igneous and metamorphic, are commonly cut with various joints. The jointing is the result of different processes such as tectonism, weathering, leaching, deformation, dislocation, *etc.* Joints are variably distributed in the rock bodies. In some areas, hard rocks are cut with joints exhibiting no apparent regular pattern, whereas in others, most of the joints are oriented in certain directions.

Joints in the rock may be either open or filled (mineralized), the fill being represented by sandy-clayey material or vein minerals such as quartz, calcite, pyrite, nickel silicates, copper ores, *etc.* Joints of tectonic origin are most commonly subjected to mineralization. The characters and extent of jointing depends also upon the composition and hardness of rocks. Harder rocks are less rapidly disintegrated; hence joints penetrate them down to smaller depths. The process of weathering much more rapidly produces joints in comparatively softer rocks. Hard metamorphic rocks often exhibit an insignificant jointing and a very low water capacity.

The hard rock formations are generally cut by differently oriented and mutually intersecting joints, the groundwater flows along both the open joints are free from the products of mechanical weathering and the joints partly or completely filled with loose sandy clayey products of weathering.

The distribution and migration of groundwater in jointed rocks are variable, being dependent on the character, origin, and size of cracks. Joints in hard rock contain both un-pressured and pressured waters. Groundwater head is commonly produced by the hydrostatic pressure in intersecting joints, some which occur in recharge area at higher altitudes, where the rainwater and the surface water infiltrates into ground, whereas the joints found at greater depths gives rise to artesian springs.

From the hydrogeological point of view, the frequency and extent of joints and fractures in the rocks are the most significant parameters imparting permeability and porosity for forming suitable groundwater reservoirs in the basaltic terrain. The secondary porosity (Joints and fractures) generally reduces with depth and hence the near surface (unconfined) aquifer system rarely, extends below 30 m depth (Agashe, 1990).

The middle and lower middle parts of the compact basalt flow are jointed (Fig. 5.2 & Fig. 5.3 of Chapter 5) illustrates the dug well in jointed compact basalt flow. These joints in the compact basalt are closely spaced linear along preferred orientations and the continuity of these joints is traceable over considerable distances. Such fracture zones, which have a discrete distribution in space and orientation, have been reported and described by various workers (Powar, 1981; Peshwa *et al,* 1987, Kale and Kulkarni, 1992 and Kulkarni, 1992). Such fracture zones transect several Deccan basaltic units and constitute recharge conduits (lateral as well as down ward transmission of water) for deeper Deccan basaltic aquifers (Kulkarni *et al.,* 1994). However, quantity of percolation of water depends upon joint spacing and pattern of jointing (Fig. 5.4). Water can percolate through closely spaced joints faster as compared to broadly spaced and non-interconnected joints. Hydrogeology of these rocks revealed that the types of basalt such as compact (massive) basalt and amygdaloidal basalt yield good amount of groundwater only when the joints, fractures and deep weathering occurs in the rock (Dhokarikar 1991; Joshi *et al,* 1996; Murthy and Jayaram 1996; Babar and Kaplay 1999; Babar 2000, Babar 2003 and Chavan *et al* 2002).

Hydrogeomorphic Expression of Lineaments

Linear features on the surface of the earth have attracted the attention of geologists for over one hundred years. This interest has grown most rapidly since the introduction of aerial photographs into geological studies. Geologists have recently proven that lineaments perceived in remotely sensed images are reliable indicators of geologic structure (Caran *et al,* 1982). A lineament is defined as a large-scale linear feature, which expresses itself in terms of topography, which is in itself, an expression of the underlying structural features. Lineaments have been used in many applications: petroleum and mineral exploration (Blanchet, 1957); nuclear energy facility siting (Seay, 1979); geothermal assessments (Woodruff *et al,* 1982); surface and subsurface feature expression in lineaments (Woodruff and Caran, 1984) and water resource investigations (Lattman and Parizek, 1964). Lattman and Parizek (1964) established a relationship between the occurrence of groundwater and fracture traces for carbonate aquifers, and in particular that lineaments are underlain by zones of localized weathering and increased permeability and porosity. La Riccia and Rauch (1977) tested this theory on a limestone aquifer in Frederick Valley, Maryland. They reported that short lineaments are usually associated with higher transmissivities, thus greater water well productivities.

Lineaments have been called lineation, linear, fracture traces, and many other names. Woodruff *et al* (1982) did a thorough review of published works on lineaments and have proposed a concise terminology, and a systematic method of perceiving and interpreting lineaments that improves data reproducibility.

Lineaments are polygenetic and inherently ambiguous features. They are ambiguous because they cannot always be field-verified, nor are they precisely reproducible. Lineaments are polygenetic in that they owe their expression to a number of different causes. Lineaments could be (a) straight stream and valley segments, (b) aligned surface sags and depressions, (c) soil tonal changes revealing variations in soil moisture, (d) alignments in vegetation, (e) vegetation type and height changes, and (f) abrupt topographic changes. All of these phenomena might be a result of structural phenomena such as faults, joint sets, or folds.

From the groundwater point of view such features may include, valleys controlled by faulting and jointing, hill ranges and ridges, displacements and abrupt truncation of rocks, straight streams and right angles off setting of stream courses etc., (Ramesh, 1990). A lineament is a regional scale linear or curvilinear feature, pattern or changes in pattern that can be identified in a data set and attributed to a geologic formation or structure (Qureshy and Hinze,1989). The linear features can be measured and treated quantitatively like measurements of other geological properties, but it is necessary to use formal statistics that reflect the circular nature of the directional data (Cheeney 1983; Gaile and Burt 1980 and Gumbel *et al* 1953; Mardia 1972). Lineaments are linear fractures commonly associated with dislocation and deformations. They provide the pathways for groundwater movement and are hydrogeologically very important Sankar *et.al* 1996. Lineaments are important in rocks where secondary permeability and porosity dominate and inter-granular characteristics combine in secondary openings influencing weathering, soil water and groundwater movements. The fracture zone forms an interlaced network of high transmissivity and serve as groundwater conduits in massive rocks in inter-fracture areas.

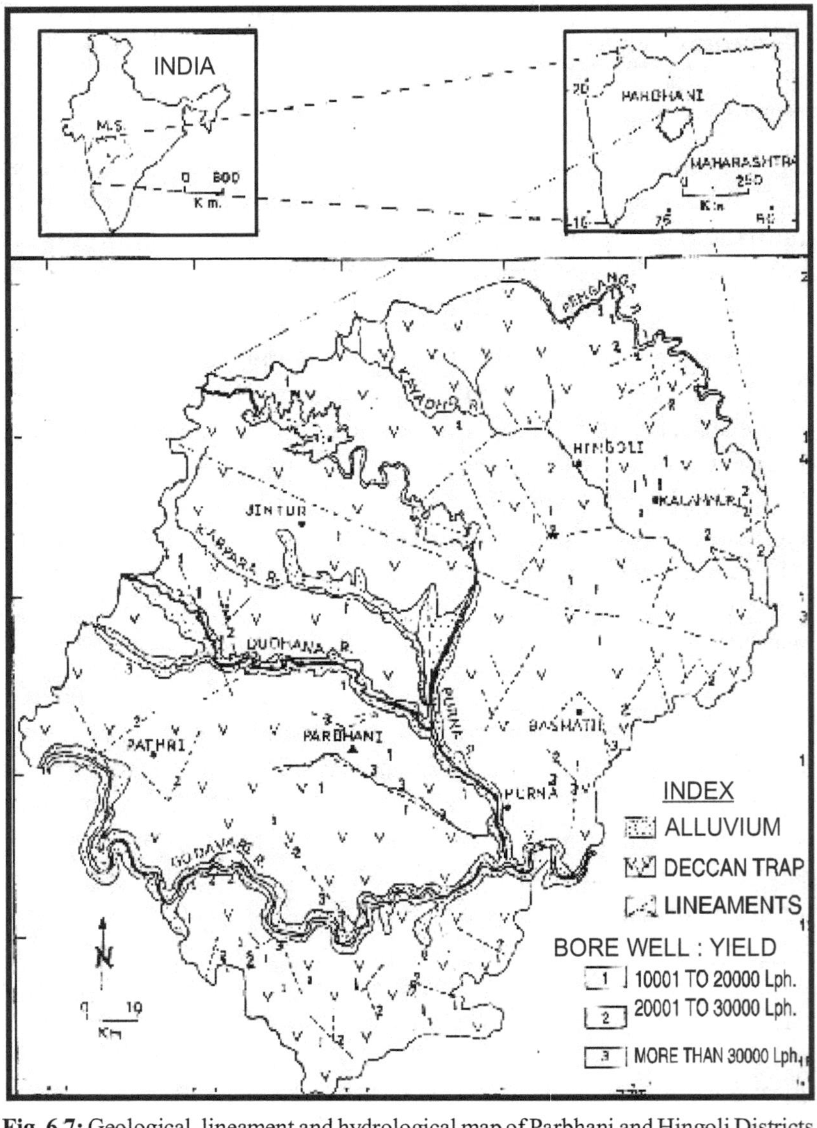

Fig. 6.7: Geological, lineament and hydrological map of Parbhani and Hingoli Districts.

In the Deccan basalt there are closely spaced linear features along preferred orientations and the continuity of these joints is traceable over considerable distances. Such linear features are called lineaments, which have a discrete distribution in space and orientation, have been reported and described by various workers (Powar 1981; Peshwa *et al,* 1987, Kale and Kulkarni 1992 and Kulkarni 1992). Such lineaments have fracture zones transecting several Deccan basaltic units and constitute recharge conduits (lateral as well as down ward transmission of water) for deeper Deccan basaltic aquifers (Kulkarni *et al.,* 1994).

The lineament intersection areas are considered as good groundwater potential zones. The combination of fractures and topographically low grounds can also serve as the best aquifers horizons (Karanth 1989; Subba Rao 1994). A set of numerous lineaments was marked in the Parbhani District (Maharashtra), India trending in the directions of north-east to south-west, north to south, a few of them east-west direction are the sites of high yielding groundwater (Murthy and Jayaram 1996, Babar 2002 a and b and Babar and Kaplay 2003) (Fig.6.7). The length of the lineaments in this area varies from a few km to several kilometers.

Chapter 7

Drainage Pattern and Morphometric Characteristics of Hydrogeomorphic Units

≈≈

⇒ This chapter deals with the hydrogeomorphic characteristics of types of drainage pattern and morphometric attributes including methods of stream orders, bifurcation ratio, length ratio, area ratio, basin shape or basin configuration, drainage density, stream frequency, slope and altrimetric analysis. All these morphometric attributes are explained with reference to the research work carried by the author.

Geomorphometric analysis is the measurement of the three-dimensional geometry of landforms and has traditionally been applied to watersheds, drainages, hill slopes, beaches, and other groupings of terrain features. In particular basin morphometric parameters have received a lot of attention from hydrologists and geomorphologists since watersheds (catchments) have been used for analysis of various physical ecosystem processes, including soil erosion, deposition, runoff, stream discharge, sediment yield, sedimentation of streams, irradiation by sunlight, evaporation, evapotranspiration, and nutrient distribution. The study of morphometric parameters, which include variables such as average basin slope, the basin elongation ratio, compactness ratio, basin relief and stream density. Some parameters, such as slope length or stream sinuosity, can be used to represent both basin and hill slope, or basin and stream channel properties.

The main task before geomorphologists is to use an ideal unit of the earth surface for the study of its landforms. The search for an ideal areal unit, within which the collection, processing, organization and interpretation of the data of the geometry of landform, particularly of erosional origin can be made has been the main aim of the geomorphologists right from Fennemen (1914), Horton (1932 and 1945), Strahler (1956 and 1957), Miller (1953), Chorley (1957), Schumm (1956), Shreve (1966), Singh (1981), Sarkar (1995), Schumm and Spitz (1996), Sharma and Amin (1996), Iqbaluddin *et al* (1997) and Raj *et al* (1999).

Basin Morphometry is a means of numerically analyzing or mathematically quantifying aspects of drainage ways/channels. Characteristics that can be measured for comparison include: stream numbers, average stream length, total stream length, stream bifurcations, basin shape, stream drainage density, basin relief, basin height, *basin area, etc.* (Singh, 1972).

Overall Basin Parameters include: basin shape, basin relief, basin area, and basin height. Drainage basins can be outlined on aerial photographs and topographic maps. Roughly the divide between drainage basins lies midway between the head-ward tributaries of two different stream systems. Area can be calculated using gridded plastic overlays or by shading the area of one drainage basin within a map of known regional dimensions (area) and digitizing for computer analysis. All these morphometric attributes are explained with reference to the research work carried by the author in Purna River and Terna River basin of Marathwada region of Maharashtra, India.

Types of Drainage Pattern

Spatial arrangement of streams giving rise to a particular design is called the drainage pattern. The drainage pattern reflects original slope, structures like folds, faults, joints and wrappings. Drainage pattern also furnish the information about the deposition by river, wind, sea, glacier and volcanic activity. Some of the basic drainage patterns (Fig. 7.1) are described below:

(i) Dendritic pattern: In the dendritic drainage pattern (Fig. 7.1a) the network of tributaries of various orders and magnitudes joining the trunk or masters stream resembles the branches of a tree. Dendritic patterns are typical of adjusted systems on erodable sediments and uniformly dipping bedrock. The development of dendritic drainage is associated with the areas of homogenous

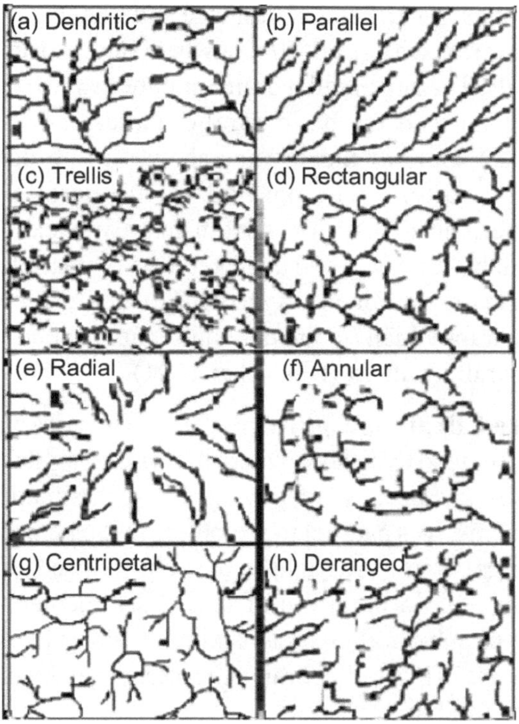

Figure 7.1: Common types of drainage patterns.

lithology and gently sloping to almost horizontal or flat topographic surfaces having extremely low relief. These types of drainage development in basaltic lava flows, which have horizontal disposition and offer nearly uniform resistance to erosion, are inferred to be the main reason for this development (Dhokrikar 1991).

Sub-dendritic pattern differs from the dendritic only in the lack perfection of development of streams. Minor deviations from dendritic pattern are presumably due to secondary regional control, either structural or topographic (Howard 1967). Fig. 7.2 shows the sub dendritic drainage pattern exhibited by some streams, which could be either due to structural or topographic control. Dendritic and sub dendritic patterns are governed by regional slope, homogenous lithology and relief (Babar and Kaplay 1999).

(ii) **Parallel pattern:** Parallel drainage patterns (Fig. 7.1b) are often found in areas with steep relief or where flow is over non-cohesive materials. The sub parallel drainage pattern shows relatively less parallelism than the basic parallel pattern (Zernitz 1932). In (Fig. 7.2) some of the streams show the sub parallel type of drainage pattern. The parallel to sub parallel drainage is characteristic of basaltic terrain mainly because of stepped topography. It has been suggested that the parallel drainage in Deccan basalt terrain is initiated due to the step like nature of the Deccan traps which is joined by subsequent lateral ravines giving a sub parallel pattern (Dhokrikar 1991).

(iii) **Trellised drainage patterns:** Trellised drainage patterns (Fig. 7.1c) tend to develop where there is strong structural control upon streams because of geology. In such situations, channels align themselves parallel to structures in the bedrock with minor tributaries coming in at right angles. The trellised drainage pattern may develop on the areas of folded sedimentary rock or as a system of sub-parallel streams aligned along the strike of rock formations. Thus trellis drainage pattern may be of structural or nonstructural origin.

(iv) **Rectangular pattern:** In rectangular drainage pattern (Fig. 7.1d) the tributary stream join the master stream almost at right angle. Even the major streams exhibit sharp right angle turn forming such drainage. Areas with intersecting tectonic faults or bedrock joints can cause streams to take on a grid-like or rectangular pattern. Rectangular pattern of drainage is developed when the lava flows are strewn with sets of joints. The joints and/or fracture in the basaltic terrain around Dudhana River may control the rectangular drainage pattern (Babar and Kaplay 1999).

(v) **Radial drainage pattern:** Radial drainage pattern (Fig. 7.1e), also known as centrifugal pattern, is formed by the streams, which diverge in all directions from a central higher point. It is developed around the denudational plateau tops, volcanoes and the dome shaped areas.

(vi) **Annular drainage pattern:** Annular drainage pattern (Fig. 7.1f) is a ring like pattern, which is subsequent in origin and is associated with maturely dissected domes and basins. Such types domes and basins are characterised by concentric circular ridges and low lands.

(vii) Centripetal drainage pattern: The centripetal drainage pattern (Fig. 7.1g) is the reverse of redial drainage pattern and occurs in the areas of karst topography exhibited by limestones, at the sinkholes. Glacial kettle-holes, volcanic craters and other depressions also exhibit the centripetal drainage pattern.

(viii) Deranged drainage patterns: Deranged drainage patterns (Fig. 7.1h) are the distorted pattern of aimlessly directed short streams, ponds and wetland areas. They are also found in areas recently disturbed by events like glacial activity or volcanic deposition. Over time, the stream will adjust the topography of such regions by transporting sediment to improve flow and channel pattern.

Besides these the special type of drainage pattern observed in the Akoli watershed area is the dendritic with radial-annular enclave pattern (Fig. 7.2). Howard (1967) suggested an anomalous type of drainage pattern in which both radial and annular patterns are combined to form the dendritic pattern. In Deccan trap area the main drainage pattern is dendritic with some minor variations. Though the main drainage pattern is dendritic but sub dendritic, radial, rectangular, sub parallel and dendritic with radial-annular enclave are also observed in few patches (Fig.7.2).

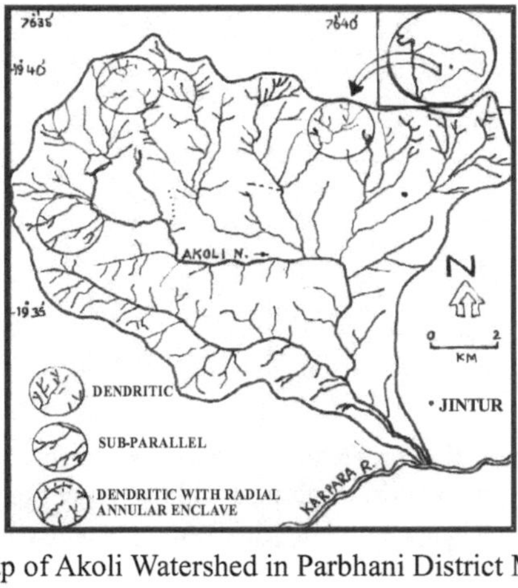

Fig.7.2: Drainage map of Akoli Watershed in Parbhani District Maharashtra.

Drainage Anomaly: Drainage anomaly may be defined as the local deviation from the regional drainage. It may include planimetric shape, pattern, depth and width of the valley, various flood plain characteristics and presence or absence of other features. The drainage anomaly helps in interpreting the structural control on the drainage development. It also provides clues to structural features and topographic conditions.

Few of the drainage anomalies are as follows:

i) ***Linear streams:*** A normal meandering stream may suddenly acquire linearity due to presence of linear features such as faults, fractures, joints, non-resistant rocks etc.

ii) ***Arcuate streams:*** The arcuate streams may be caused due to domal up-warps or singly plunging folds. Parallel arcuate drainage is typically exhibited by singly plunging anticline (Fig 7.3a), while doubly plunging anticlines exhibits opposite parallel arcs in the drainage (Fig 7.3b).

iii) ***Radial drainage*** is associated with domes (Fig. 7. 1).

iv) ***Lithological variations*** cause change in drainage density e.g. Sandstones show low drainage density with coarser texture, while shale shows higher drainage density with finer texture (Fig 7.3c). Off set drainage is usually found in the area with the faults controlling the drainage pattern (Fig. 7.3d).

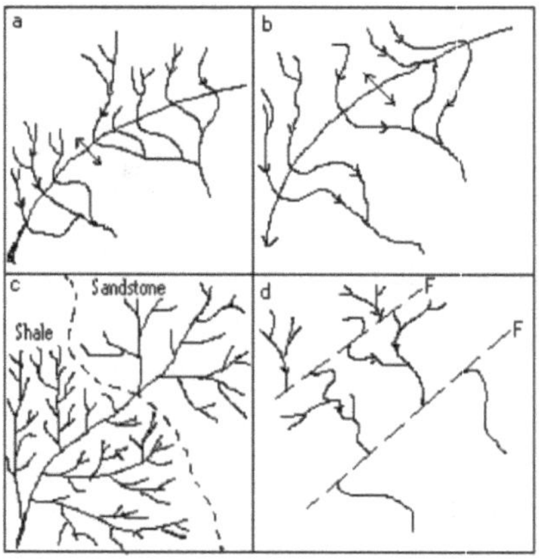

Fig. 7.3: Drainage Anomalies (a) Parallel arcuate drainage in singly plunging anticline (b) Opposite parallel arcuate drainage in doubly plunging anticline (c) Change in drainage density due to change in lithology (d) Abrupt change in stream direction due to presence of fault.

The possible evidence of structural control on drainage include sharp bends in streams, sudden right angled turn of streams, streams flowing parallel to major rivers for a considerable distances, convergence and divergence of streams and in general anomalous behaviour of streams (Fig. 7.4 a to l).

The streams encircled in Fig. 7.4 (a, c, h, i, j, k and l) show the sudden bend or turn before joining the Terna River. This indicates that the stream courses are structurally controlled (Babar *et al,* 2000).

Two streams encircled in Fig. 7.4 (b, f and g) show the convergence and divergence behaviour. The streams should have a confluence with one another before joining the Terna River but have diverted and taken the sudden turn.

The streams encircled in Fig.7.4 (c, d, e and h) instead of joining the Terna River through shortest route followed the course of it for considerable distance.

These anomalous behaviours of streams are some of the evidences of structural or tectonic control on the region.

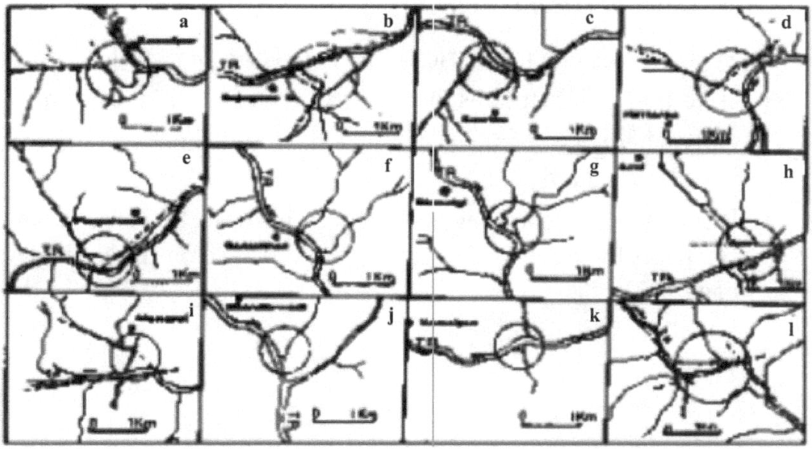

-------------- Major Lineaments — Macro Lineaments interepreted on the basis of drainage pattern
.... Probable part of the stream which should have been there under normal conditions. T.R. Terna River.

Fig. 7.4: Anomalous drainage pattern along Terna River. In a, c, d, e, j, k- streams flowing parallel to River for some distances in the upstream direction and then join indicating the structural control over drainage; b, f, g- streams converge and diverge before joining the River; h, i, l- the sudden turn and shift of streams along the course of Terna River.

Methods of Stream Orders

Measurements of the numbers of streams, length of streams, branching of streams, etc. may be used to numerically/ statistically quantify the nature of the stream pattern and be used as a predictive tool to stream behavior primarily in discharge. For example streams with very high branching patterns usually spread runoff more evenly across the landscape and more evenly to streams, resulting in lower and usually slower discharges per stream segment. Streams with a low amount of branching usually funnel water to the few stream segments that do exist and result in high discharges. Branching can be numerically analysed by counting the number of stream segments and calculating ratios between stream segments (primary tributaries vs. secondary tributaries and so on).

Stream segments are assigned numbers based upon their position within the river tributary system. Assignments of segment numerical values have been developed in several different ways, the most commonly used are:

Strahler Method: (Used) **(Fig.7.5)**

- Segment with no tributaries is designed as a first-order stream
- Two first order streams join to form a second-order stream
- Only where two segments of equal magnitude join is an increase in order required
- Bifurcation ratios are calculated: by dividing the first order stream segments by the second order segments and so on (Described on page 117).

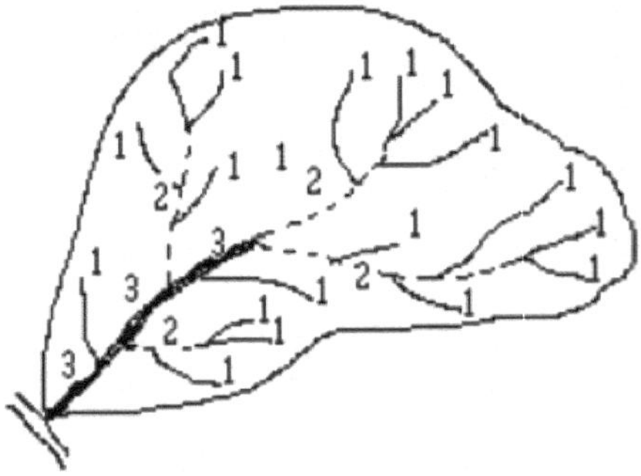

Fig. 7.5: Stream ordering after Strahler, (1957)

Shreve Method: probably commonly used

- Segment with no tributaries is designed as a first-order stream
- Additional stream links represent the sum of the link numbers of all tributaries that feed it.
- Bifurcation ratios are calculated: such as the first order segments divided by the of second order segments (and so on)

Bifurcation Ratios, Length Ratio and Area Ratio

Bifurcation Ratios: Stream Bifurcation Patterns or branching patterns of streams vary depending upon bedrock control. Where uniform materials underlie the drainage basin, the streams usually branch randomly. Where folds or faults control weakness and stream development, the development of branches can be restricted to zones of faults or fracture or weak sedimentary strata.

Out of the four system of ordering the streams that are available (Gravelius 1914, Horton 1945, Strahler 1957 and Scheidegger, 1970), the system of Strahler (1957) modified after Horton (1945) is commonly used because of its simplicity. Accordingly the smallest un-branched stream segment was designated as the first order stream, the one formed by the merging of two such first order segments was designated as the second order stream and so on. In this ordering, the Karpara river, Parbhani, Maharashtra (India) was found to be a sixth order stream, Dhamuda a fifth order stream and Pingalgad a fourth order stream (Fig. 7.6, Babar and Kaplay 1999).

Bifurcation ratio is related to the branching pattern of the drainage network. It is expressed in terms of the following equation

$$Rb = \frac{N}{N+1}$$

Where,

N = Number of streams of a given order

N +1 = Number of streams of the next higher order.

The bifurcation ratios of three sub-basins Purna River basin are given in Table 7.1 (Babar and Kaplay 1999).

A- Karpara Sub-basin B-Dhamuda Sub-basin C-Pingalgad Sub-basin

Fig. 7.6: Drainage map of Purna River Basin Parbhani, India.

Branching pattern analysis is most commonly examined by counting the number of tributaries of the same rank or same level of contribution.

Significance/implications of Bifurcation ratios:

- High bifurcation ratios are common in areas with structural control of drainage ways (folded valley and ridge regions; faults).

  High bifurcation ratios also are common in areas of "mature dendritic" drainage patterns.
- High bifurcation ratios tend to signify streams that have a higher than average flood potential (Eash 1996) because numerous tributary segments drain into relatively few trunk transporting stream segments.
- High bifurcation ratios include ratio values reaching upward of 20:1 and more.
- Low bifurcation ratios are more common in areas with uniform surficial materials
- Where geology is reasonably homogenous, ratios usually range from 3.0 to 5.0.

- Low bifurcation ratios commonly signify areas with high drainage density and therefore fewer collecting segments per transporting segment (sometime check out Strahler or one of the others and look a bifurcation values and their implications)

The bifurcation ratios of three sub basins of Purna River basin India (Babar and Kaplay, 1999) are given in Table 7.1. During this study the bifurcation ratios of Karpara sub-basin are found to be ranging from 2.00 to 4.69, Dhamuda ranges from 2.00 to 3.91 and that of Pingalgad sub-basin ranges from 3.72 to 10.00 (Table 7.1). Bifurcation ratios of streams of these three sub basins range from 2.00 to 5.30 indicating that there is apparently minimal structural control in drainage development. Bifurcation ratio of 10 for the third and fourth order stream of Pingalgad sub-basin is an exception to this. Such abnormally high bifurcation ratio may be due to elongated shape of the basin (Strahler, 1964 and Zavoianu, 1985).

Table 7.1: Bifurcation ratios of three sub-basins of Purna River.

Stream Order	**Karpara Sub-basin**		**Dhamuda Sub-basin**		**Pingalgad Sub-basin**	
	N	R_b	N	R_b	N	R_b
1	401		86		197	
		3.57		3.91		3.72
2	136		22		53	
		4.69		3.14		5.30
3	29		7		10	
		3.22		3.50		10.00
4	9		2		1	
		4.50		2.00		
5	2		1			
		2.00				
6	1					

N= Number of Streams R_b= Bifurcation Ratios

The number of streams decreases as the stream order increases. The relation of number of streams against stream order in each of the three sub basin (Fig. 7.7) shows that the number of streams of a given order forms an inverse geometric sequence by decreasing systematically with increasing order and conforms to the Horton's (1945) law of stream numbers. Horton (1945) law of stream number states that 'the numbers of stream segments of each order form an inverse geometric sequence with numbers of stream order'. Karpara, Dhamuda and Pingalgad sub basins have negative correlation between stream number and stream order. The correlation coefficient (r) values for Karpara, Dhamuda and Pingalgad sub basins are -0.797, -0.837 and –0.898 respectively. The corresponding values for goodness of a fit (r^2) are 0.635, 0.701 and 0.806 respectively (Fig.7.7).

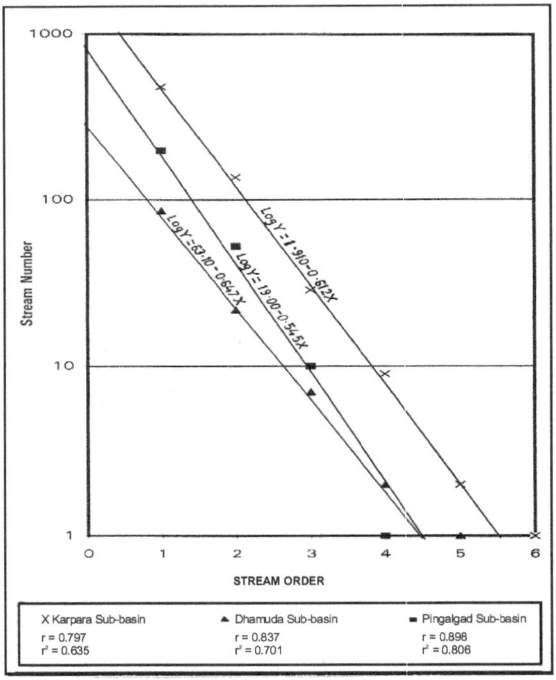

Fig. 7.7: Plot of stream order against stream numbers of three sub basins.

Stream Length: Stream length involves measurement of the length of the different rank/order streams is a drainage system. Average stream lengths can be compared mathematically to total drainage area, basin relief, *etc.* Several techniques for measuring stream length involve computer digitization of separate stream segments of the same order. In general, with trellis large-scale drainage patterns have high ratio values of length *vs.* other parameters (such as length *vs.* drainage basin area or length *vs*. bifurcation ratio). High values of length to other parameter ratios usually correspond to flash discharge behavior of stream systems.

A number of authors have published findings on generalities with respect to morphometric analyses and stream systems and their underlying geology. Ritter in his 2002 textbook and Baker *et al* 1988 have generalized that:

- Areas with trellis patterns tend to have very high bifurcation ratios and flashy discharge patterns.
- Areas with dense dendritic, rectangular, radial, and parallel patterns tend to have low bifurcation ratios and non-flashy discharge patterns and are usually developed in unconsolidated or softer surficial materials.
- Areas with low density dendritic patterns tend to have high bifurcation ratios and flashy discharge patterns – usually in less erodable consolidated bedrock
- Areas with large basins in humid settings with sandstones, shale, and/or mixed sedimentary strata tend to have low stream densities and therefore tend to be flashy
- Small carbonate basins in sub-arid settings have very low stream densities (probably is telling us much drainage is through karst) but normally not an intensely flashy discharge pattern because much of the drainage is by groundwater flow which results in lower surface runoff and lower discharger.

Lengths of each stream segment of each order are measured for each sub-basin. The total and mean lengths and the length ratios calculated for three sub-basins are given in table 7.2

The values of mean stream length for the area are in direct geometric progression to stream order (Fig.7.8) and conforms to the Horton's (1945) law of stream length. The law states that 'The average lengths of streams of each of the different orders in a drainage basin tend to approximate a direct geometric series in which the first term is average length of streams of 1st order'.

Table 7.2: Length ratios of three sub-basins of Purna River.

Stream Order	Karpara Sub-basin			Dhamuda Sub-basin			Pingalgad Sub-basin		
	L	L_m	R_l	L	L_m	R_l	L	L_m	R_l
1	316.0	0.68		77.5	0.90		167.4	0.85	
			1.35			1.32			1.62
2	122.2	0.91		26.3	1.19		73.2	1.38	
			3.25			2.85			1.87
3	92.0	2.97		23.8	3.40		25.8	2.58	
			2.22			1.29			16.43
4	52.6	6.58		8.8	4.38		42.4	42.4	
			1.08			0.86			
5	14.3	7.14		3.8	3.80				
			1.04						
6	7.4	7.40							

L= Length of Streams L_m= Mean Stream Length R_l= Length Ratio

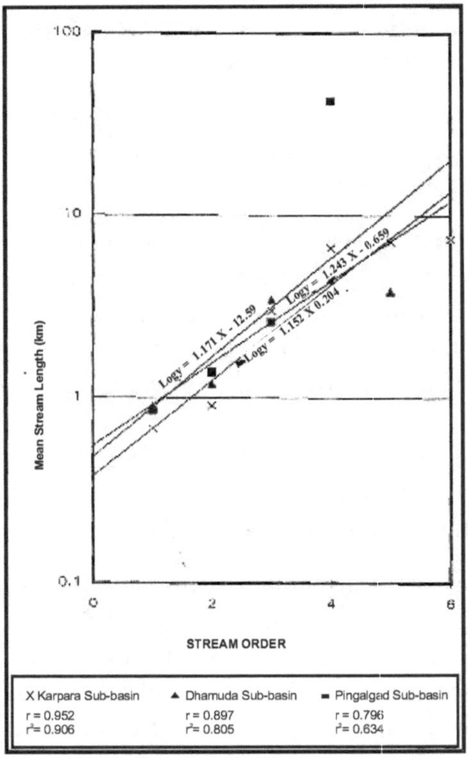

Fig. 7.8: Plot of stream order against mean stream length of three sub basins.

The length of first order streams of Karpara and Pingalgad sub basins are, however, slightly more than the average trend (Table 7.2). The regression equations of stream length to stream order are identical for all the sub basins. The correlation coefficient (r) and goodness of fit (r^2) for mean stream length and stream order for Karpara, Dhamuda and Pingalgad sub basins are $r = 0.952$, $r^2 = 0.906$, $r = 0.897$, $r^2 = 0.805$ and $r = 0.796$, $r^2 = 0.634$ respectively.

Stream Area Ratios: The area of a drainage basin is an important parameter like the length of the stream draining it. The areas of various drainage basins of different orders have been planimetrically (computer area curvimeter) measured and from these the mean area of the basins of each order is computed. Total stream areas of each order and stream area ratio of corresponding pair of streams are given in (Table 7.3. Fig. 7.9) shows the plots of the mean stream area against the stream order.

Table 7.3: Area ratios of three sub-basins of Purna River.

Stream Order	Karpara Sub-basin			Dhamuda Sub-basin			Pingalgad Sub-basin		
	A	A_m	R_a	A	A_m	R_a	A	A_m	R_a
1	271.40	0.56		59.94	0.70		135.2	0.69	
			3.92			4.92			4.89
2	297.65	2.19		75.44	3.43		177.6	3.35	
			6.29			5.26			3.00
3	399.02	13.76		126.2	18.04		100.5	10.05	
			1.86			3.92			31.1
4	229.85	25.54		141.4	70.72		312.4	312.4	
			7.95			2.29			
5	405.83	202.92		161.9	161.9				
			2.28						
6	462.15	462.15							

A= Area of Streams A_m= Mean Stream Area R_l= Area Ratio

The plot of mean stream area versus corresponding stream order (Fig.7.9) shows that, the mean basin area is in direct geometric progression, which is in conformity with the Schumm's (1956) law of mean basin area. The law states that 'The mean drainage basin area of streams of each of the different orders in a drainage basin tend to approximate a direct geometric series in which the first term is mean area of streams of 1st order'. The values of correlation coefficient (r) and goodness of fit (r^2) for these parameters are higher for Dhamuda sub basin (r = 0.904, r^2 0.818) followed by Karpara sub basin (r= 0.841, r^2 = 0.708) and are lowest for Pingalgad sub basin (r = 0.790, r^2 = 0.624).

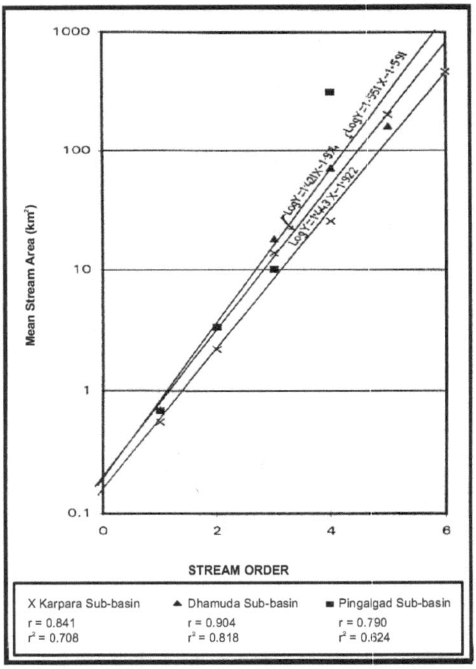

Fig. 7.9: Plot of stream order against mean basin area of three sub basins.

Basin Shape or Basin Configuration

The basin configuration is important as it helps in the description and comparison of different forms of the drainage basin. It is related to the functioning of the units of the basin and also it's genesis. Different popular methods of computation of basin shape are as follows:

i) Form factor (F): The form factor suggested by Strahler (1968) is the ratio of the basin area to the squared value of the basin length (L)

$F = A / L^2$

Where F = Form factor, A = Basin area and L = Basin length

The form factor value varies from 0 (in highly elongated shape) to the unity *i.e.* 1 (in perfect circular shape). Hence, higher the value of form factor more circular the shape of the basin and vice-versa.

The form factor obtained for three sub-basins are Karpara = 0.33, Dhamuda = 0.23 and Pingalgad = 0.17 (Table 7.4).

ii) Elipticity Index (E): Elipticity index is calculated using the method of Stoddart (1965) as follows:

$E = \pi L^2 / 4A$

Where, E = Elipticity index, A = Basin area and L = Basin length.

The value of elipticity index E varies from 1 to infinity and it is inversely proportional to 'F'. The values of 'E' obtained are 2.38 for Karpara, 3.41 for Dhamuda and 4.62 for Pingalgad sub-basins (Table 7.4).

ii) Circularity ratio (C): The data on the circularity ratio is given in Table 7.2 and it is calculated using the formula:

$C = 4\pi A / P^2$

Where, P = basin perimeter.

The values of 'C' vary from 0 (in a line) to 1 (in a circle). Higher the value of circularity ratio more circular shape of the basin and vice-versa.

The values of 'C' thus obtained are 0.57 for Karpara, 0.55 for Dhamuda and 0.40 for Pingalgad sub basin (Table 7.4).

Elongation ratio (R): Elongation ratio is the ratio of diameter of the circle with the same area as basin and the length of the basin.

$$R = \frac{\textit{Diameter of the circle with the same area}}{\textit{Basin length}}$$

$$R = \frac{2\sqrt{A/\pi}}{L} = \frac{2}{\pi}\sqrt{\frac{A}{L^2}}$$

$$R = 2\sqrt{F / \sqrt{\pi}}$$

or

$$F = \pi / 4R^2$$

This means that Schumm's elongation ratio 'R' is proportional to the square root of Hortons form factor 'F'. The value of R varies from 0 (in highly elongated shape) to unity *i.e.* 1.0 (in circular shape). Thus higher the value of elongation ratio more circular shape of the basin and vice-versa. The values of elongation ratios obtained are Karpara = 0.65, Dhamoda = 0.54 and Pingalgad = 0.46 (Table 7.4).

The four parameters of basin shape, *viz.* Form factor, elipticity index, circularity ratio and elongation ratio of three sub- basins of Purna River basin show that all of them are greatest for Karpara and Dhamuda sub basins signifying that they are more circular sub basins, whereas, the lowest values for Pingalgad sub basin indicate it to be the elongated sub basin. The lower values for Pingalgad sub basin may be due to presence of fracture lineament and jointed sub surface rock.

Drainage Density and Stream Frequency

Drainage Density: Drainage density is total length of stream channel versus the entire area of the drainage basin. High drainage densities usually reduce the discharge in any single stream, more evenly distributing runoff and speeding runoff into secondary and tertiary streams. Usually there is little basin lag between rainfall initiation and rise of water in all stream segments. Usually there is a moderate rise followed by a decline that usually stretches over several days. An area with high drainage density means that it takes water a while for water to drain to a primary stream and time to arrive at secondary streams; so that there is basin lag. This "lag" time between rainfall and corresponding discharge is referred to as "Basin Lag". Basin lag is a natural means of "regulating" the flow of water through a system and effectively "slows" flooding downstream. Flooding still occurs down stream, but there is a "time buffer" before floodwaters crest in downstream portions of stream systems. Hydrographs of basins with dense stream channels are more broadly rounded and hydrographs of larger basins and "downstream" portions of basins tend to be more broadly rounded in shape. Where drainage density is very low, intense rainfall events are more likely to result in high discharge to a few streams and therefore a greater likelihood of "flashy" discharge and flooding in humid areas of low drainage densities (such as 3-4 miles per square mile) suggest resistant bedrock in humid areas; high drainage densities (such as hundreds of miles per sq. mile) suggest highly erodable surficial materials.

Measurement/Determination of Stream Density: One of the easier ways is to obtain aerial photographs of small scale (7 1/2 minute quadrangle scale of 1:24,000 is probably on the high side — if aerial surveys of 1-to-several thousand exist; they are most useful). The entire area of the drainage basin may be shaded. Streams may be outlined on a separate map that can be digitized as well and the total stream length calculated. Density is calculated:

$$Drainage\ density = \frac{Total\ stream\ length}{Total\ drainage\ basin\ area}$$

Dd = L / A,

Where, L = Total lengths of all stream segments of a basin, A = Total area of the basin.

The Karpara sub basin is found to have a drainage density of 1.31 km/km^2, dhamuda sub basin has Dd of 0.87 km/km^2 and Pingalgad sub basin had Dd of 0.99 km/km^2 (Table 7.4).

Stream Frequency (Fs)

The stream frequency can be derived by using formula:

F = N / A,

Where, N = Total number of stream, A = Total area of the basin.

The values of stream frequency obtained for Karpara, Dhamuda and Pingalgad sub basins are 1.43, 0.73 and 0.84 respectively (Table 7.4). All these values are expressed as streams/km^2.

Regions of highly resistant or highly permeable sub soil, under vegetation on the other hand favour development of lower drainage

Table 7.4: Basin configuration of three Sub-basins of Purna River.

Sr. No.	Parameters	Karpara Sub-basin	Dhamuda Sub-basin	Pingalgad Sub-basin
1	Total Basin area (A) Sq.Km	462.15	161.88	312.38
2	Max. Basin length (l) Km	37.5	26.8	43.0
3	Form Factor (A/l^2)	0.33	0.23	0.17
4	Basin Perimeter (P) Km	101.33	60.78	99.60
5	Radius of the circle with P as the circumference Km	16.14	9.68	15.86
6	Area of circle with same perimeter (A_0) Sq. Km	817.97	294.23	789.83
7	Circularity Ratio A/A_0	0.57	0.55	0.40
8	Compactness Ratio K=4ÐA/P^2	0.57	0.55	0.40
9	Diameter of the circle with area as a basin (l_0) Km	24.26	14.36	19.94
10	Elongation Ratio (l_0/l)	0.65	0.54	0.46
11	Total stream length (L) Km	604.5	140.2	308.8
12	Drainage Density (L/A) Km/km^2	1.31	0.87	0.99
13	Total Number of streams (N)	663	118	261
14	Stream Frequency (Fs=N/A) Streams/Km^2	1.43	0.73	0.84
15	Length of overland flow $1/2D_d$ Km.	0.38	0.68	0.60

density (<1 km/km^2) and low stream frequency (<1 stream/km^2) (Strahler, 1964). In the area of study the drainage density and stream frequency are somewhat higher for Karpara sub basin (Dd = 1.31 km/km^2 and Fs = 1.43 streams/km^2, Table 7.4), which correspond to the rocky, hilly terrain of the sub basin with sparse vegetation and impermeable sub surface strata.

The value of drainage density and stream frequency are quite lower for Dhamuda sub basin (Dd = 0.87 km/km^2 and Fs = 0.73 streams/km^2) and Pingalgad sub basin (Dd = 0.99 km/km^2 and Fs = 0.84 streams/km^2, Table 7.4), indicative of the permeable soils with high moisture retentive capacity and relatively low relief of these sub basins.

Length of overland flow (lg): This term refers to the length of the run of the rainwater on the ground surface before it gets localised into definite channels (Horton 1945). The length of overland flow is considered as a dominant hydrologic and morphometric factor and is the mean horizontal length of flow path from the water divide to the stream in a first order basin and is a measure of stream spacing and degree of dissection, Chorley (1969). Lengths of overland flow thus calculated for Karpara, Dhamuda and Pingalgad sub basins are 0.38 km, 0.68 km and 0.60 km respectively (Table7.4).

The streams of Purna River in alluvial zone show higher values of length of overland flow (0.60 and 0.68 streams / km^2 for Pingalgad and Dhamuda sub basins respectively) compared to the streams flowing in upland zone (0.38 streams / km^2 for Karpara sub basin).

Stream Channel evolution through time: Stream channels and the systems they comprise obviously change through time, but there are differing opinions on how stream systems change, the pattern of change, the rate of change, etc. Because stream erosion is slow by human standards, the changes are mostly theoretical. Changes in stream patterns can be postulated from field study/examination of relict stream structures (palaeo stream structures) and flume analyses based on stream table studies (stream tables that are small scale simulations of rivers). From relict stream studies and flume simulations, some very broad generalizations regarding changes in basin morphometry through time can perhaps be made. One such change pattern is diagrammed in Easterbrook based on research by Parker in 1976. Parker used materials to simulate uniform consolidated surficial materials. In the stream table simulation Parker used the initially developed a dense dendritic pattern such would be expected in a temperate area with uniform surface material of low slope. Streams and therefore their patterns grow by head-ward extension until divides are reached.

In Parker's simulation, streams did, in fact, grow by head-ward extension. During simulated extension phases, drainage density increased, especially around the outside margins of the basin. During the extension phases drainage density decreased in the lower (downstream or "center) portion of the basin as a consequence of loss of channels.

Essentially the high density dendritic pattern in the center of the basin "evolved" into a low density dendritic pattern with time; so the discharge pattern would be expected to change – discharge patterns would probably become more "flashy" with evolution of the drainage basin. That change in discharge patterns with evolution of stream patterns might be generally extrapolated — as trellis patterns age, perhaps the extension of streams to divides results in even less density of streams in the "center" portions of trellis systems and a shift to even "flashier" discharge. It may be stretching the study, but it is arguable that the center of most drainage basins tends toward flashier discharge as they age.

Sinuosity Index: Sinuosity deals with the pattern of channel of a drainage basin. Sinuosity index value greater than 1.5 are characteristics of meandering course of a river (Richards 1982 and Singh 1998).

The sinuosity indices explain the hydrological and topological characteristics of drainage basin. Thus channel sinuosity is an important parameter used in the morphometric studies to distinguish between the various types of terrain and to ascertain the degree of establishment made by the drainage line in the area of influence.

Any tectonic deformation that changes the slope of a river valley will result in corresponding changes in sinuosity so as to maintain an equilibrium channel slope (Keller and Pinter, 1996). Thus sinuosity parameters can be used to deduce the role of tectonism development of channel morphology (*eg.* Gomez and Marron 1991;Rhea 1993; Rachna Raj, 1999). The sinuosity of a meandering stream is the result of topography and hydraulic factors, which can be expressed by a ratio called the index of sinuosity (Miller 1968). The river channel is divided into number of segments (Fig. 7.10) as suggested by Muller (1968) and Friend and Sinha (1988) for determination of sinuosity parameter. The measurements of channel length (CL), valley length (VL) and the shortest distance between the source and the mouth of the river (AL) i.e. air lengths are used for calculation of channel index, CI=CL/AL and valley index VI= VL/AL.

The standard sinuosity index is calculated using SSI= CL/VL, hydrological sinuosity index HIS= %equivalence of CI-VI/CI-1 and topographical sinuosity index TSI=% equivalent of VI-1/CI-1. The standard sinuosity index (SSI) for the Terna River channel varies from 1.002 to 1.76 (Table 7.5). The increase in SSI in the lower reaches of the basin is accompanied by lower values of Hydraulic Sinuosity Index (HSI), which suggests that the hydraulic factor is not responsible for increase in SSI in the lower reaches. The values of HSI vary from 2.13 to 85.58 (Table 7.5).

Table 7.5: Sinuosity variation in Terna River channel

Seg. No.	CL (Km)	VL (Km)	AL (Km)	CI	VI	SSI	HSI	TSI
1	8.60	8.52	8.36	1.033	1.020	1.014	39.39	60.61
2	9.72	9.70	8.52	1.141	1.138	1.002	2. 13	97.87
3	6.48	6.35	5.76	1.125	1.102	1.020	18.40	81.60
4	5.28	5.04	4.68	1.128	1.077	1.048	39.84	60.16
5	7.32	6.97	6.25	1.171	1.115	1.050	32.75	67.25
6	6.49	6.44	5.76	1.127	1.118	1.008	56.52	43.48
7	5.76	5.54	5.08	1.134	1.091	1.040	32.75	67.25
8	6.00	5.78	5.40	1.111	1.087	1.022	21.62	78.38
9	5.64	5.62	5.40	1.044	1.041	1.004	6.82	93.18
10	7.49	6.94	6.25	1.198	1.110	1.079	44.44	55.56
11	9.72	9.24	8.43	1.153	1.096	1.052	37.25	62.75
12	8.88	8.16	8.04	1.104	1.015	1.088	85.58	14.42
13	7.08	6.27	6.09	1.163	1.030	1.129	81.60	18.40
14	6.24	5.48	4.08	1.529	1.343	1.138	35.16	64.84
15	8.16	7.23	6.25	1.306	1.157	1.129	48.69	51.31
16	6.61	5.66	5.16	1.281	1.097	1.168	65.48	34.52
17	6.27	5.60	5.25	1.194	1.067	1.119	65.45	34.55
18	7.32	6.25	5.52	1.326	1.132	1.171	59.51	40.49
19	5.28	4.49	4.20	1.255	1.069	1.176	69.33	30.67
20	4.44	3.79	3.46	1.284	1.094	1.174	66.90	33.10

CL-Channel length, VL- Valley length, AL-Air length, CI-Channel index, VI- Valley index SSI-Standard Sinuosity Index, HSI- Hydraulic sinuosity index, TSI- Topographic sinuosity index

Slope and altimetric analysis

Basin Relief is the difference in elevation between the highest head-ward tributaries and the mouth of the main drainage stream. Basin height is simply the relief of the basin above a chosen level. In general,

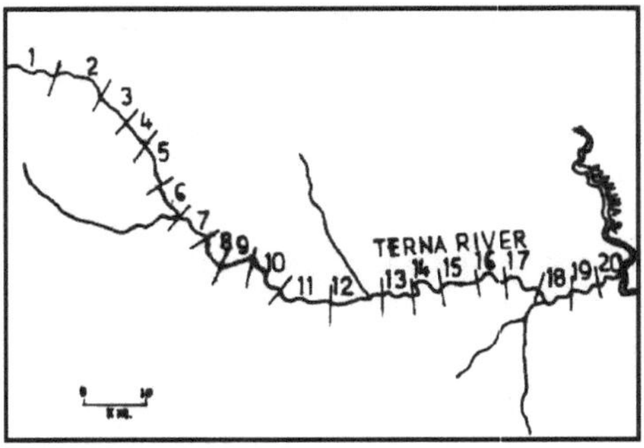

Fig. 7.10: Map showing segments studied for sinuosity. Segments for sinuosity.

the larger the basin the more likely it is to have basin runoff lag. In general, the higher the relief, the shorter the time for precipitation/ basin runoff lag; so overall basin parameters do give some indication of stream behavior and predictability of that behavior.

Slope aspects and altitude are important terrain parameters from land utilization point of view and also in assessing disaster prone zones. Among the three parameters of the terrain (slope, aspects and altitude), slope is very important for assessing land capability, erodibility, stability and irrigability. However, for landuse all the three aspects may be necessary.

The relief aspects of the drainage basins are related to the study of three-dimensional features of the basin involving area, volume and altitude of vertical dimensions of landforms. These terrain characteristics are the results of basin processes. Different morphometric methods are used to analyse these features. These methods include the analysis of the channel gradient, longitudinal profile, clinographic curve, slope, maximum basin relief, relief ratio, ruggedness number and relationship between area and altitude (hypsometric analysis).

a) ***Channel gradient:*** Channel gradient can be calculated from the total drops in elevation from the source to the mouth and the horizontal distances between these two points are measured along their channels. The respective gradients computed for Karpara, Dhamuda and Pingalgad streams streams are found to be 2.54 m/km, 2.17 m/km and 1.51 m/km respectively. The data is presented in Table 7.6 and their longitudinal profiles are given in Fig 7.11.

b) ***Clinographic analysis:*** The clinographic curves reveal the breaks in slope and sudden changes in the relief of the area and they also present the general trend of the surface. Thus the slope angle between two successive contours is calculated using the techniques suggested by Strahler (1952). For this formula given below is used.

Tan Q = CI/AW, where CI = Contour interval, AW = Average width between two successive contours calculated as A/(L1+L2)/2, where, A is total area between contours and L1 and L2 are lengths of two successive contours.

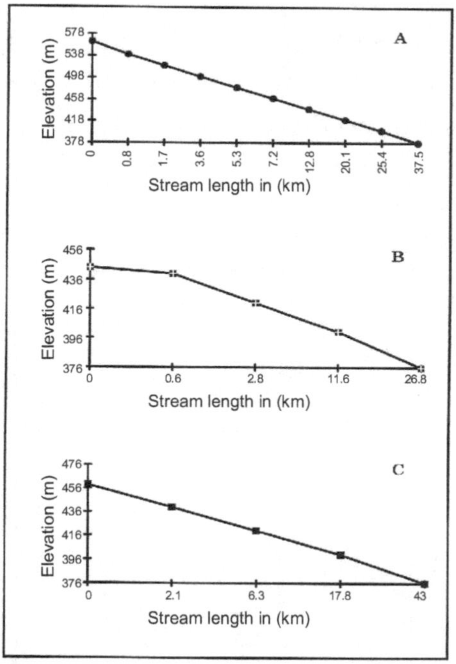

Fig. 7.11: Longitudinal profiles of three sub basins.

The data of the mean slope is given in the table 7.6. The clinographic curve is drawn by plotting the ground slope against the contour height (elevation) as given in Fig.7.12.

Longitudinal profile and clinographic curves are of great importance to the geomorphologists who are primarily concerned with the analysis of relief and surfaces of terrain because profiles provide a visual perception of the actual nature of terrain. The clinographic and long profile curves for Dhamuda and Pingalgad sub basin are smooth concave indicating the mature state of terrain development, whereas the break in slope of profile of Karpara River (Fig.7.11-7.12) indicates nick points are elevated scraps.

Table 7.6: Relief parameters of three Sub-basins of Purna River.

S. No.	Parameters	Karpara Sub-basin	Dhamuda Sub-basin	Pingalgad Sub-basin
1	Highest point on basin perimeter m.	564	444	459
2	Depth of basin mouth m.	379	379	373
3	Maximum basin relief (H) m	185	65	86
4	Max. Basin length (l) Km	37.5	26.8	43.0
5	Relief ratio (Rh=H/l)	0.005	0.002	0.002
6	Ruggedness number (HXD_d)	0.24	0.06	0.09
7	Slope of ground surface ($Sg=H X 2D_d$)	0.48	0.12	0.18
8	Channel gradient (m/km)	2.54	2.17	1.51

D_d =Drainage Density

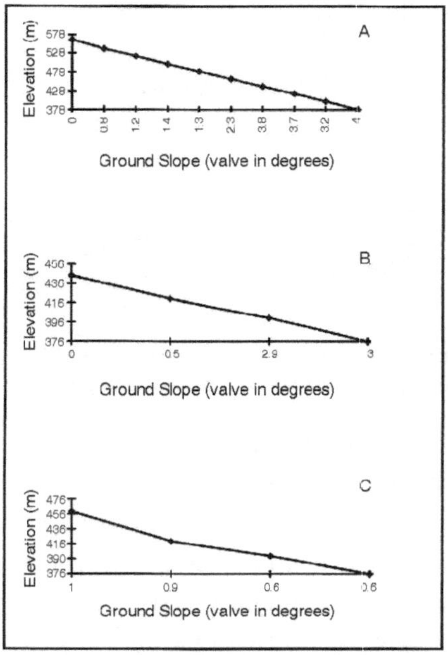

Fig. 7.12: Clinographic curves of three sub-basins.

c) ***Slope:*** Slope is the combined result of many factors like geological structure, absolute and relative relief, climate, vegetation cover, drainage texture and frequency, *etc*. It is one of the significant morphometric attributes in the study of landforms of a drainage basin (Singh and Srivastava 1974). Computation of slope angles from field measurements involves tedious and time-consuming procedures. Several techniques of the derivation and computation of average slopes from topographical maps have been suggested by various workers (Rich 1916; Wentworth, 1930; Smith, 1938; Raise and Henery 1937; Calef 1950; Smith 1950; Strahler 1956; Miller and Sammersoo, 1960; Eyles, 1965; Schumm and Mosley 1973). In the present study the technique of Wentworth (1930), which is easier and involves lesser measurement and calculation and more rapid procedure than other schemes has been used.

In Wentworth's method the contour map of the region is divided into 1 km x 1 km grid and the numbers of contour crossings in these grids are counted. Actual slope angle is computed from the following formula:

Slope angle = $\tan Q = N \times I / 636.6$ (MKS system)

Where,

N = number of contour crossings per km lengths and I = Contour interval in meters.

The values of slope angles derived for each grid square of the drainage basin are classified into convenient slope categories.

Different categories of the slope are:

i) Level slope (Sl) = 0 – 2
ii) Gentle slope (SG) = 2 – 5
iii) Moderate slope (SMS) = 5 – 15
iv) Steep slope (SS) = > 15

The slope map (Fig.7.12) reflects that greater than 90% of the area of Dhamuda and Pingalgad sub basins belongs to slope category of gently sloping to nearly level terrain (0 – 5 slope). The slope map of

Karpara sub basin (Fig.7.13) reveals that the slope group (0-2) covers 18.78%, (2- 5) covers 29.26%, (5-10) covers 31.02%, (10-15) covers 16.34% and (>15) covers 4.60% area of the total area of the sub basin. The Karpara is the only sub basin dominated by moderate to steep slope category indicating that it belongs to the hilly terrain.

Aspects and slope have direct bearing on vegetation type and conditions. All India Soil and Land Use Survey (AIS&LUS) has prepared guidelines for preparation of slope categories which can be safely adopted for deriving slope classes on 1:50,000 scale. The slope categories that can be derived from Survey of India toposheets on the guidelines of AIS&LUS are shown below:

Table 7.7: Slope characteristics of geomorphic surfaces

S. No.	Slope categories	Slope (%)
1.	Nearly level	0 - 1
2.	Very gently sloping	1 - 3
3.	Gently sloping	3 - 5
4.	Moderate sloping	5 - 10
5.	Strong sloping	10 - 15
6.	Moderately steep to steep sloping	15 - 35
7.	Very steep sloping	> 35

d) ***Maximum basin relief (H):*** The maximum basin relief is obtained from the elevation difference between highest point on the basin perimeter and its confluence with the trunk stream. The values of maximum basin relief obtained for Karpara, Dhamuda and Pingalgad sub basins are 185 m, 65 m and 86 m respectively (Table 7.6)

e) ***Relief ratio (RL):*** When basin relief H is divided by the horizontal distance on which it is measured, there result a dimensionless ratio called relief ratio. The relief ratios obtained for Karpara, Dhamuda and Pingalgad sub basins are 0.005, 0.002 and 0.002 respectively (Table 7.6). The relief ratio measures the overall steepness of the drainage basin and is an indicator of the intensity of the erosion process operating slopes of the basin. Possibility

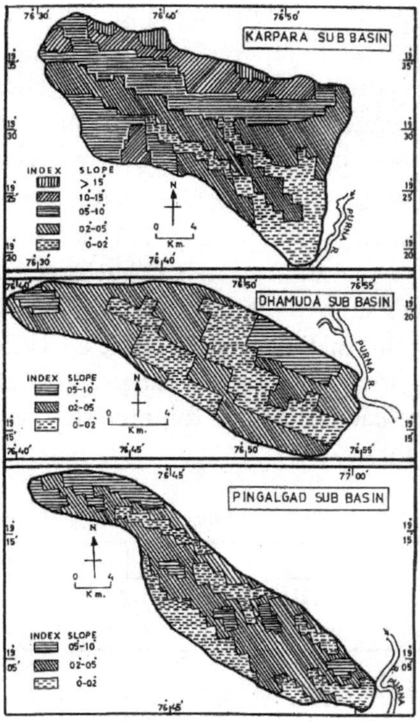

Fig. 7.13: Slope map of three sub basins.

of the close relation between relief ratio and hydrologic characteristics of a basin is suggested by Schumm (1954), who found that the sediment loss per unit area is closely correlated with the relief ratio.

f) ***Ruggedness number:*** Ruggedness number is the product of basin relief (H) and drainage density (Dd), where both terms must be in same units. It is a dimension less quantity used to combine the qualities of slope, steepness and length. Generally the ruggedness values range from as low as 0.06 in subdued low relief (Plain areas) to over 1.0 in mountain ranges or in badlands or on weak clays. The ruggedness numbers calculated for Karpara, Dhamuda and Pingalgad sub basins are 0.24, 0.06 and 0.09 respectively (Table 7.6).

The relief measures for Karpara sub basin are relatively higher with maximum basin relief of 185 m, relief ratio of 0.005 and ruggedness number of 0.24, all indicating that the basin lies within hilly terrain. The relief measures for Dhamuda and Pingalgad sub basins are relatively lower *i.e.* maximum basin relief (65m and 86m), relative relief (0.11 and 0.09), relief ratio (0.002 and 0.002) and ruggedness number (0.06 and 0.09) indicating that these sub basins lie within flat or gently sloping terrain.

g) **Hypsometric (area-altitude) Analysis:** The hypsometric curve has been termed the drainage basin relief graph (Vladimirescu, 1978). Langbein (1947) appears to have been the first one to use such a line of study to collect hydrologic data. Strahler (1952) popularized this method. Hypsometric analysis is the relation of horizontal cross sectional drainage basin area to elevation. Hypsometric data calculated for three sub-basins of Purna river is given in Table 7.8.

Table 7.8: Hypsometric data of three sub-basins of Purna River

Contour (m)	a	a/A	h	h/H
KARPARA SUB-BASIN [A=462.15Sq.Km H=185 m]				
560	2.85	0.006	182	0.98
540	14.19	0.018	162	0.88
520	24.58	0.099	142	0.77
500	38.99	0.106	122	0.66
480	60.26	0.109	102	0.55
460	92.16	0.199	82	0.44
440	178.06	0.385	62	0.34
420	266.10	0.424	42	0.23
400	370.62	0.802	22	0.12
380	461.45	0.998	02	0.01
DHAMUDA SUB-BASIN [A=161.88Sq.Km H=65 m]				
440	5.84	0.036	61	0.94
420	47.13	0.291	41	0.63
400	152.69	0.943	21	0.32
380	161.48	0.997	1	0.02
PINGALGAD SUB-BASIN [A=312.38Sq.Km H=86 m]				
440	5.48	0.016	65	0.76
420	45.86	0.147	45	0.52
400	158.85	0.509	25	0.29
380	279.22	0.894	5	0.06

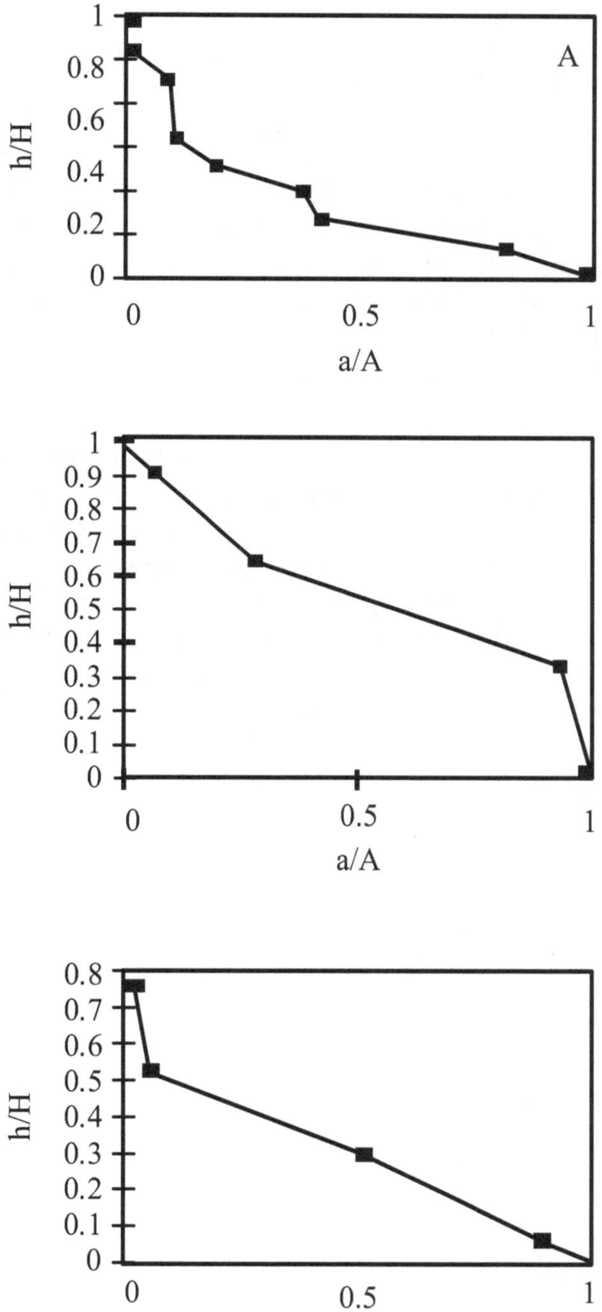

Fig. 7.1 4: Hypsometric curves for three sub basins.

In the present study of (percentage hypsometric integral method) two ratios are computed from map measurements of contour map (Fig. 7.14). First one is a/A where, 'a' is the area enclosed between a given contour within the basin and basin boundary and A is the total area of the basin. The second ratio is h/H, where, h is the height of the contour above the base level of the stream mouth and H is the total height of the basin with reference to the same base (Table 7.8). A graph of a/A versus h/H is plotted for each sub basin (Fig.7.14). The area below the respective curves is measured and its proportion (in percent) with reference to the total area of the graph calculated. The hypsometric integrals thus obtained for Karpara, Dhamuda and Pingalgad sub basins are 31, 57 and 32 percent respectively (Fig.7.14).

The values of the hypsometric integrals and the form of the hypsometric curves taken together help to identify the stage of the basin development (Strahler, 1952). These integrals in study area range from 31 to 57 %, which indicates that the sub basins are in early mature stage of erosional development. Hypsometric curve is useful in hydrologic applications, where some hydrologic factors such as precipitation or evaporation vary with altitude or where the vegetation cover shows altitude stratification.

Chapter 8

Hydrogeomorphology of Landforms

≈≈≈

⇒ This chapter deals with the hydrogeomorphological studies of erosional, depositional, denudational, and structural landforms.

The beginning of the 20th century was heralded by methodological revolution in geomorphological studies brought in by Davis (1899). His classical model of geographical cycle defined by him ' as a period of time during which and uplifted landmass undergoes its transformation by the process of land sculpture ending into a low featureless plain (peneplain)' dominated the geomorphological investigations all over the world throughout 1st half of the 20th century.

The statistical techniques by Horton (1932 and 1945) brought quantitative revolution in the field of geomorphology when he presented quantitative analysis of morphometric characteristics of fluvially originated drainage basin. Post 1950 geomorphology has undergone vast change in the methods and approaches to the study of landforms, conceptual framework, paradigm and thrust areas of study. The most outstanding contribution to the advancement of geomorphological knowledge in this period is the adoption of quantitative approach based on deductive scientific method to the study of landforms and processes at short spatial and temporal scales (Strahler, 1956 and Schumm, 1956).

Several attempts have been made in the world in past to classify large territories and landforms into morpho-units. Valuable contributions pertaining to landform analysis have been summarized below: Joerg (1914) subdivided North America into natural regions, Fenneman (1914) established physiographic boundaries and gave physiographic divisions of the United States, whereas Anderson (1941) gave physiographic divisions of Columbia plateau. Howard and Spock (1940) and Weaver (1965) presented the excellent ways of classification of the landforms while Wood and Snell (1960) applied the quantitative systems for the classification of the landforms. However, Linton (1951) and Hammond (1954 and 1965) gave techniques for mapping delineating small scale and large-scale landforms.

Important works related to applied aspects of landform analysis and classification of large territories and valuable contributors in geomorphology have made landforms into morpho-units in various parts of India. They are: Singh (1969)-evolution of Chotanagpur Highlands; Singh (1978) – Physiographic regions, landforms and erosion surfaces of Ranchi plateau; Bhattacharya and Banerjee (1979) – Quaternary geology and geomorphology of Ajay-Bhagirathy valley; Vaidyanadhan (1983) – landforms studies in Peninsular India; Singh (1983) - the landscape cycles of Peninsular India; Nathawat (1984) – applied remote sensing data for geomorphological and environmental studies; Singh (1985) - systematic approach to the study of landforms; Singh and Rastogi (1992) – morphoagricultural regions of Belan-Son; Singh and Rastogi (1994) – morphodynamics and morphogenesis in sub-humid tropical environment; Singh (1995) – geomorphology in the appraisal of natural resources and Goswami *et al* (1996) – geomorphological mapping of Assam using remote sensing techniques.

The significant contributions in geomorphology of Deccan basalt terrain is made by Rajaguru (1969) on late Pleistocene of the Deccan; Dikshit (1970) – on erosion surfaces and polycyclic reliefs of Deccan trap and Kale (1983) – on geomorphological repercussion on Indian Peninsular movement and Western Ghat formations.

Fluvial Landforms

Landforms shaped by running water are described as fluvial landforms and are shaped by fluvial processes, such as overland flow and stream flow. Fluvial landforms dominate the continental land surfaces, and are heavily exploited for agriculture. Fluvial landforms can be further divided into two groups: (1) erosional landforms are sequential landforms shaped by progressive removal of particles, and (2) depositional landforms created by the accumulation of transported sediments that have been removed from the sequential landforms.

All streams carry sediment. The actual amount is related to the characteristics of the materials on which the stream flows. Usually, an increase in discharge results in an increase in the amount of sediment that is carried. A sediment rating curve models the relationship between discharge and the amount of sediment carried. Now within the stream sediment can be carried in three ways: bed load; suspended load and dissolved load.

One of the most obvious landforms created by streams are the channels they flow in. We can classify stream channels into three types. Channels located in the upper reach of streams are V-shaped, narrow, and deep with little associated floodplain. These channels are often cut into bedrock and limited bed sediments consist loose rocks and boulders. Stream channels become shallow, wide, and braided when the stream gradient changes from being steep to gently sloping. Shallow braided channels are created by the presence of sand and gravel beds in a floodplain. These sediments are deposited because a sudden change in gradients causes a reduction in flow velocity. Often a drastic change in grade also causes the deposition of an alluvial fan. Channels become U-shaped and meandering with continued reduction in stream gradient. Meandering channels flow over an extensive floodplain of complex deposits.

We can describe a number of features within the stream channel. Streams carrying coarse loads develop sand and gravel beds when velocity is reduced spatially or temporally. Point bars are common in meandering streams developing on the inside of channel bends. In straight streams, bar deposits form in relation to the thalweg. Some other features associated with the thalweg include riffles and pools. Some other features common within the stream channel are dunes and ripples. These features form on the channel bottom when the bed is composed of sand and silt.

Floodplains are flat areas adjacent to stream channel that are composed primarily of deposits put down during episodes of floods. A number of features are found on the floodplain. Some of these features are depositional like levees, abandoned point bars, and depressions. Other features like crevasses and Oxbow lakes are created by erosion.

When a stream ends its flow into a lake or ocean the sediment it carries will be deposited to create a delta. Deltas are complex features made up of 3 different bed types. Deltas also have a variety of different shapes.

Hydrogeomorphologists often view streams as being part of drainage basins. Drainage basins are arbitrarily defined by topography. Drainage basins are a very useful geomorphic concept and they are often modeled as open systems. Within drainage basin inputs and factors like topography, soil type, bedrock type, climate, and vegetation influence a number of stream properties. One stream property that is affected by these factors is the drainage pattern of streams.

Stream morphometry is all about exploring the mathematical relationships between various stream attributes. Studies have discovered that these relationships can be viewed as laws because of their geometric or mathematical predictability. These measurements can also be used to compare streams and basins to identify factors that may be causing differences.

Coastal Landforms

Landforms of coastal regions are in most cases shaped by the geomorphic action of ocean waves. Ocean was can exert a significant amount of energy through the kinetic energy of wave motion. The energy of ocean waves on the terrestrial landscape of the Earth is concentrated at the shoreline. At the shoreline the energy wave motion causes erosion and the transport of sediment. Erosion along the shoreline is normally greatest at areas where wave refraction causes the intensification of wave energy. Deposition occurs along the shoreline when wave energy is reduced in it intensity and when sediment is transported into an area by beach drift and long-shore drift.

Glacial Landforms

During the last billion years of Earth history there have been several periods where global average temperatures were much colder than they are today. The colder climates lead to the formation and expansion of extensive continental and alpine glaciers. The last period of glacial advance began about 2 million years ago. For the past 14,000 years we have been experiencing a warming of the global climate, which has lead to a retreat of glaciers. Today glaciers cover about 10 % of the

Earth's land surface. During the height of the last glacial advance about 30 % of the Earth's land surface was covered. Geomorphologists have classified glaciers based on size.

The growth of glaciers first involves the conversion of snow into glacial ice. This process takes many years and the weight of overlying deposits to covert snow with a density between 50 to 300 kilograms per cubic meter to ice with a density of 850 kilograms per cubic meter. For mass of ice to be classified as a glacier it most have the ability to move. This occurs when the mass of ice becomes so heavy that it begins to flow by plastic deformation. Masses of ice in mountain valleys begin to flow when accumulations become more than 20 meters high. Rates of flows within glaciers are influenced by factors like friction, ice weight, and slope of the valley. Some glaciers are able to move more quickly because they experience basal sliding.

Hydrogeomorphologists often view glaciers as systems that are influenced by a number of inputs and outputs. Some of these inputs and outputs influence the glaciers ability to surge or retreat. The concept of glacier mass balance suggests that two processes influence a glacier: accumulation and ablation. If accumulation exceeds ablation the glacier surges forward. If ablation exceeds accumulation the glacier surges retreats.

Glaciers have played an important role in shaping the landscape of the middle and higher latitudes. The effects of glaciers exist at essentially two scales. Alpine glaciers modify landscapes on a local scale. Many alpine glaciers are still operating today in the world's mountainous regions. Continental glaciers influenced landscapes on the regional and continental scale. Most of these glaciers are now gone, but the evidence of their action is still quite obvious.

Glacial landforms that are the result of erosion can once again occur at two scales like local and large scales. On the local scale, alpine glaciers, for example are responsible for the erosional features such as: U-shaped valleys, hanging valleys, cirques, horns, and aretes. Large-scale landforms in the continental glaciers remove large amount of surface sediments on the areas where they occurred. Sometimes this removal of material left large depression, which can be filled with lakes of various sizes. Both types of glaciers leave the following depositional features: till, moraines, outwash plains, and erratics. The size of these deposits is of course much large for the continental glaciers. Continental glaciers also produce a number unique deposits like: kames, eskers, and drumlins.

Periglacial processes influence about 25 % of the Earth's terrestrial surface. The most important process of periglacial environments is freeze-thaw action. Extreme cold temperatures of periglacial environments often cause surface soils and sediments to freeze. These frozen sediments are called permafrost. Many types of permafrost have been classified. Permafrost can have a depth of up to 1500 meters. On its surface, is a layer that is subject to seasonal thawing. This layer is called the active layer. Sheets of permafrost can have unfrozen layers on top, beneath, or within it. These unfrozen layers are called taliks.

Periglacial areas are also known for the presence of ground ice. Ground ice comes in a variety of different forms and each type has unique characteristics and processes of formation. A number of geomorphic processes operate in periglacial regions. One process that is quite active in periglacial regions is physical weathering due to the crystallization of water. This process can create vast quantities of shattered angular rock on the periglacial landscape. Mass movement is common in periglacial regions of the world. The main processes that are resulting into mass movement include solifluction, gelifluction, frost creep, and rock falls. Finally, erosion occurs in periglacial environment by way of invention, aeolian processes, and fluvial processes.

Three types of landforms are common in periglacial environments. The first is patterned ground. Pattern ground comes in a variety of shapes including stripes, steps, circles, polygons, and nets. A number of processes are involved in the formation of these similar features. Periglacial environments are also characterized by the presence of ice mounds. Palsas are low elongated mounds with cores of segregated ice and peat. The other type of ice-cored hill is called a pingo. These features are much larger than palsas and are formed by either cryostatic pressure or artesian groundwater flow.

Aeolian Landforms

Places in the arid regions and coastal areas of the world are influence by wind. The speed of these winds is great enough to move soil particles that range in size from clay to sand. Material is transport by wind by three processes: suspension, saltation and traction. The finest particles as clay and silt are carried in suspension for longer time for greater heights and longer distances before they are washed down by rain. The coarser material of sand grade is carried by saltation. These particles are lifted up by the wind but do not remain in suspension for long. They are dropped down as the velocity decreases and picked up again when the velocity increases. Thus such particles travel by leaps and jumps, which is known as saltation. The saltating particles, which are moving with nearly the same speed as the wind, strikes the large particles with force proportional to their size and velocity. These impacts may cause the large particles to be pushed slowly in the wind direction. The movement is known as traction.

The power of wind produces a variety of aeolian erosional features. Some of these features include deflation hollows, pans, yardangs, mushroom shaped rocks, rock columns, pinnacles, ventifacts and desert pavement. Wind also produces a variety of depositional features. Sand dune is the most noticeable feature of deposition. Sand dunes come in a variety of forms that are produced by a range of processes. Another important deposit of aeolian forces is loess. Many deposits of loess were formed in the past when winds blew over glacial deposits during the Pleistocene.

Classification of Landforms

In general the landforms can be classified as erosional, depositional, denudational and structural. Hydrological characteristics, composition of surface materials, soil conditions, drainage and vegetation pattern, depth of weathered materials which have specific recharging, storing and transmitting capacity are discussed in this chapter. This would help in knowing the nature and water potentiality of different geomorphic units and landforms for identification of its ground water prospect zone rating from excellent to poor. The groundwater prospects or various hydrogeomorphic units and associated landforms can be identified and described in brief as under.

Most of the recent studies in the geomorphology are concentrated towards fluvially originated geomorphic surfaces in the world right from oldest floodplain deposits to the recently developed landforms.

Some of the notable studies are: Dumont *et al,* 1996; Hattingh 1996; Mertes 1994; Mertes *et al,* 1996; Schumm *et al,* 1996; Kraus 1997; Lecee 1997; Leeder, 1997; McFadden and McAulifce 1997; Twidale 1997; Konrad 1998; Lekach 1998 and Zhang 1998.

The geomorphic surfaces are classified into different morpho-units (Landforms) on the basis of physiographic characteristics, morphological features, relief, slope, drainage density and lithology. These landforms are described below with reference to the hydrological point of view:

Erosional Landforms

The landforms resulting from varied fluvial erosional processes include river valleys, waterfalls, potholes, river terraces, meanders, dissected plateaus, mesa and buttes and peneplains, *etc*.

River Valleys

The development of river valley starts from *splash erosion* by raindrops. It is a very effective method that causes soil erosion. During a torrential downpour, up to 100 tons of soil per acre can be dislodged by splash erosion. *Sheet erosion* occurs when overland flow removes soil in uniform thin layers, often resulting in a denuded farm field. Sheet erosion may occur when rainwater completely fills all pore spaces in the soil (saturated conditions) causing enhanced runoff. In areas with a steep slope, runoff from torrential rains can cause *rill erosion*. Sometimes these rills can develop into *gullies*, steep-walled trenches whose upper end grown progressively upslope. Gullies further develop into long V-shaped valleys (Fig. 8.1).

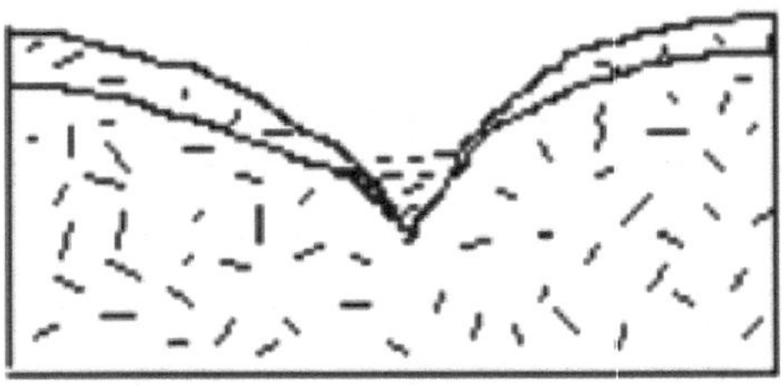

V. Shaped Valley Profile

Fig. 8.1: Development of V-shaped valley.

Most of the rivers are associated with the valley of their own. A valley may be defined as lowland surrounded on sides by various inclined hill slopes and mountains. River valleys are responsible for the origin, development and modification of their valleys through well-established processes of river erosion. The shape and dimensions of fluvially originated valleys change with the advancement of the stages of fluvial cycle of erosion. The valley formed in the youthful stage of fluvial cycle of erosion and in the initial stage valley development is V-shaped having steep valley side slope of convex element. Valley channel located in the upper reaches of many streams tend to be narrow with flow moving at high velocities. The high flow velocities found in these streams are the result of a steep grade and gravity. Within these stream systems, erosion is a very active process as the channel tries to adjust itself to the topography of the landscape. The valley is very deep and narrow, such type of V-shaped valleys are the result of accelerated rate of down cutting *i.e.* vertical erosion or valley deepening. These valleys after continued sideward erosion are transformed into very broad and shallow valleys having concave valley side slope of very gentle gradient during old stage. V-shaped valleys are divided into two types *viz.* (i) gorges and (ii) canyons (Fig. 3.l).

The gorges and canyons represent very deep and narrow valleys having very steep valley side slopes like vertical wall. Normally, a very deep and narrow valley is called a gorge and the extended form of a gorge is called a canyon. Gorges are formed due to active down cutting of valleys through the mechanism of pothole drilling (Fig. 8.2) during the youth stage of the fluvial cycle of erosion. Gorges are also formed due to the recession of waterfalls (Fig. 8.3). Most of the Himalayan Rivers have carved deep and narrow gorges. Nighoj potholes of Kukadi River (Fig. 8.2) and Sahastrakund waterfall on Painganga River in Maharashtra, India (Fig. 8.3) show the formation of Gorges. The Grand Canyon of the Colorado River in the state of Arizona (USA) having length of 482.8 kilometers and a depth of 2088.3 m is one of the most important canyons of the world. The Indus River has cut across the Himalayan ranges and flows through 17,000 ft deep gorge and canyon.

Fig. 8.2: Gorge developed due to the potholes on Kukadi River at Nighoj, Maharashtra.

Fig. 8.3: Gorge due to waterfall at Sahastrakund waterfall on Painganga River in Maharashtra.

Waterfalls

Waterfalls are defined as the magnificent jumps made by streams at different parts of their course where there is a sudden and considerable drop in the gradient of river channel. A waterfall may also be defined as a vertical drop of water of enormous volume from a greater height in the long profiles of the rivers. Rapids are of much smaller dimensions than waterfall. In waterfalls the stream water falls from a height and acquires normal flow again some distance below.

Waterfalls are caused because of sudden descents or abrupt breaks in the longitudinal course of the rivers due to a host factor *e.g.* variation in the relative resistance of rocks, relative difference in topographic reliefs, fall in the sea level and related rejuvenation, earth movements, *etc*. There are numerous examples of well known waterfalls developed in the rivers of the world. In India many famous waterfalls are located. The best examples are Gersoppa fall on the Sharavati River, Siva smudran falls of Cauvery River in Mysore and the waterfall on Penganga River located at Sahastrakund in Nanded District Maharashtra, India (Fig. 8.4).

Fig. 8.4: The waterfall on Penganga River located at Sahastrakund in Nanded District Maharashtra

Potholes

Potholes results due to diverted water currents that creates eddies where water swirls round and round with great speed. The load of river spin along with eddies and becomes churning tools. Because of this churning action the bedrock gets eroded and excavated into cylindrical holes. These holes are known as pothole. The best example of pothole formation is the potholes in the bed of Kukdi River a tributary of Ghod River (Fig. 8.5) occupying the area of about 300 sq. meters with the depth of about 20 meters. Due to churning action created by eddies, in the bed of the Kukadi River, the pebbles caught up in the smaller depressions and due to the tremendous velocity of the river water the individual potholes get widened and ultimately intersected each other. Due intersections these potholes the riverbed is converted in to the deep channel or gorges. At places several small potholes have united to form a sort of mesh.

The riverbed contains hard, dense, fresh and unjointed amygdaloidal basalt therefore the potholes in the channel are very much smooth and polished due to abrasive and polishing action of the caught up pebbles. On the sides of channel line the vertical cut thin irregular flows are clearly exposed. As there is much elevation difference between the main river and the tributary and also the tributary Kukdi river meet at right angle, the pothole are much developed. These observations suggest that the fresh amygdaloidal basalt suitable for carving and polishing.

Fig. 8.5: The potholes in the bed of Kukdi River a tributary of Ghod River, Maharashtra, India.

River Terraces

These are bench like ledges or flat surfaces that occur on the sides of many stream valleys. The narrow flat surfaces on either sides of the valley floor are called river terraces (Fig. 8.6). They represent the level of former (older) floodplains. Sometimes several terraces on either side where they are arranged in step-like forms frequent the river valleys. River terraces are generally formed due to dissection of fluvial sediments of floodplains deposited along a valley floor.

The rivers form extensive floodplains during late mature stage and attain their graded curves of profile of equilibrium. The floodplains

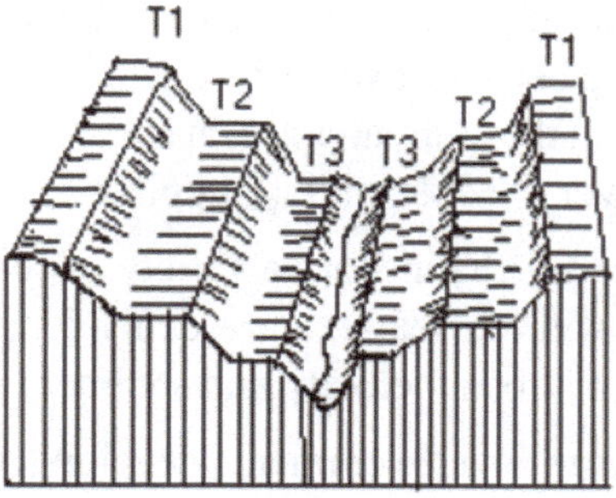

Fig. 8.6: Development of River Terraces.

consist of thick deposition of alluvium and gravels. The rivers are rejuvenated due to sudden negative change in sea level. Consequently, the erosive capacity of the rivers is increased substantially. Thus rejuvenated rivers deepen their valleys due to accelerated rate of vertical erosion. Now the rivers form their new narrow valleys within former flat valleys and thus terraces are formed on both the sides. Thus the river terraces develop in succession over the earlier formed flat valleys.

Meanders

Meander channels form where streams are flowing over a relatively flat landscape with a broad floodplain (Fig. 8.7). Technically, a stream

Fig. 8.7: Dudhana river showing meandering course, older alluvial plain and point bar deposits at Sambar, Parbhani District, India.

is said to be meandering when the ratio of actual channel length to the straight-line distance between two points on the stream channel is greater than 1.5. Channels in these streams are characteristically U-shaped and actively migrate over the extensive floodplain.

Within the stream channel are a variety of sedimentary beds and structures. Many of these features are dependent upon the complex interaction between stream velocity and sediment size.

Streams carrying coarse sediments develop sand and gravel beds. These types of bars seen often in braided streams, which are common in elevated areas. Bars develop in braided streams because of reductions in discharge. Two conditions often cause the reduction in discharge: reduction in the gradient of the stream and/or the reduction of flow after a precipitation event or spring melting of snow and ice.

Point bar develop where stream flow is locally reduced because of friction and reduced water depth. In a meandering stream, point bars tend to be common on the inside of a channel bend.

Dissected Plateau (DPT)

An extensive flat top and steep slopes formed over horizontally layered Vindhyan rocks or Deccan Basalts may be criss-crossed by fractures/Joints/lineaments etc. are called plateaus. Ground water prospects are good moderate especially at lineament intersections. Deep valleys/gullies developed due to stream/river erosion on plateau and may be criss-crossed by lineaments. These rocks have generally poor prospect of groundwater but moderate at intersection of lineaments.

(i) Highly dissected plateau (HDP)

The land of this unit is severely dissected by the streams of giving size to a terrain consisting of flat-topped ridges and steep scarps. These units can be expressed with reference to the evaluation range, total percentage of the area, morphometric attributes like slope, runoff characteristics, drainage density, stream frequency and relative relief percentage. Groundwater potential in this unit is very poor. The runoff water can be arrested in the form of check dams and other suitable measures can be taken to boost the pasture, farm forestry and horticulture in these areas. These types of landforms are very well seen in Deccan Trap region.

(ii) Moderately dissected plateau (MDP)

This geomorphic unit has widest spatial distribution in the Deccan Trap region. The main dissecting agents here are various rivers and their tributaries in Deccan Trap region. These units can be expressed with reference to the evaluation range, total percentage of the area, morphometric attributes like slope, runoff characteristics, drainage density, stream frequency and relative relief percentage. The soils covering these plateau unit are moderately thick, dark brown to black in colour clayey, calcareous, moderate to well drained and high in moisture retentive capacity. Groundwater potential in this unit is moderate to high.

(iii) Undissected plateau

The land of this unit is dissected by the streams of giving rise to undissected terrain consisting of flat-topped hills and steep scarps. These units can be expressed with reference to the elevation range, total percentage of the area, morphometric attributes like slope, runoff characteristics, drainage density, stream frequency and relative relief percentage. Groundwater potential in this unit is very poor. The runoff water can be arrested in the form of check dams and other suitable measures can be taken to boost the pasture, hill top forestry and horticulture in these areas.

Mesa (Ms) and Buttes (Bt)

These are the erosional features made up essentially of horizontally layered rocks, having a cap of hard and resistant rock that has escaped erosion. A mesa (Fig. 8.8) is generally produced in such horizontal layers of alternating characteristics that are exposed to river erosion. The small sized isolated patches of resistant horizontal layers covering the strata below are called buttes (Fig. 8.8). These flat-topped isolated plateaus with steep sided slope at the top has poor ground water prospect.

Peneplains

Peneplains represent low featureless plain having undulating surface and remnants of convexo-concave residual hills. These are, in fact, the end products of normal cycle of erosion. These are frequented with low residual hills known as monadnocks, which are left out due to

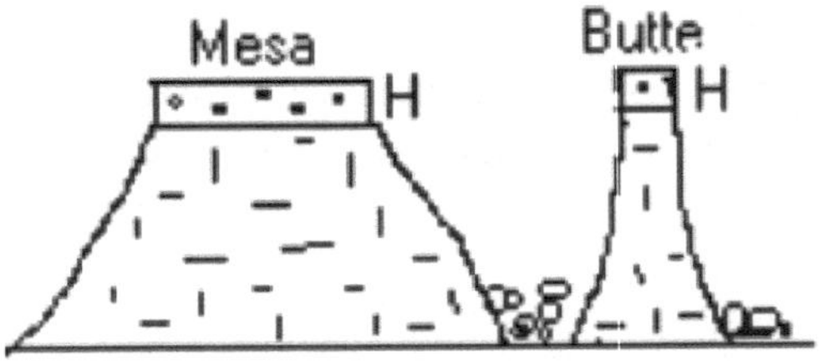

Fig.8.8: Mesa and Butes.

less erosion of relatively resistant rocks. The end product of normal or fluvial cycle of erosion has been variously named different geomorphologists such as peneplain (Davis 1899), endrumpf (Penck), panplain (Crickmay 1933), etchplain (Pugh 1966), panfan (Lawson), *etc.*

Landforms of the Coastline (Marine)

The most important agents shaping the coastal landforms are waves, which are visible forms of energy produced by any of several processes, such as winds. Waves move with little loss of energy until they reach shallow areas, when the wave steepens and forms breakers. When these waves break against a weak or soft rock type, a marine scarp or marine cliff is produced.

Continuous erosion on these marine cliffs may result in various landforms as follows:

Wave-cut notch

An erosional indentation at the base of the cliff is called Wave-cut notch (Fig. 8.9)

Sea arch

Sea arch occurs where a more resistant rock type projects out into the ocean itself (Fig. 8.9).

Sea stack

An isolated rock column that has become detached from the marine cliff is called sea stack (Fig. 8.9).

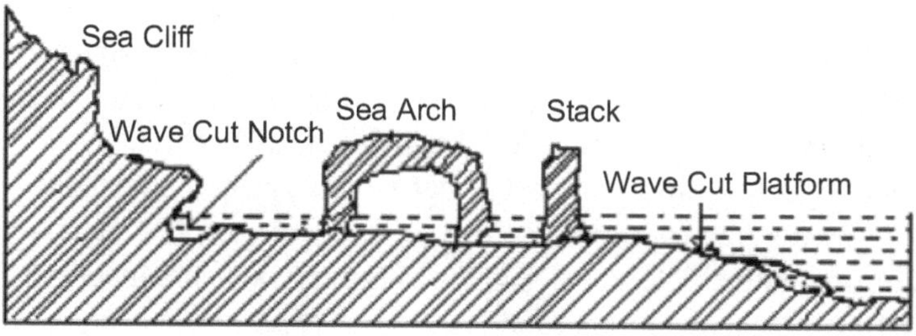

Fig. 8.9: Coastal erosional landforms.

Abrasion platform or wave cut platform

A flat, broad area at the base of the cliff consisting of more resistant rock is called abrasion platform or wave cut platform (Fig. 8.9).

Ria coast

A ria coast is formed by the submergence of a continental landmass. It may also be formed by a rise in sea level, inundating an area that has been dissected by streams. This situation creates many offshore islands that had previously been carved by streams.

Fiord coast

A fiord coast is similar to the ria coast in that it was formed by submergence of a landmass or rising sea level, but the area submerged had previously been carved by valley glaciers. Hence, the fiord coast has few beaches due to the depth of the fiord.

Depositional Landforms

Soil particles that are removed by overland flow and deposited in low-lying areas forms colluvium. Soil particles that are picked up by streams, carried downstream, and later deposited around the stream forms alluvium. Various landforms of depositional origin such as floodplain, alluvial plains, alluvial fan, river sand, ravines, meander scar, point bar deposits, delta *etc*.

Floodplain (FP)

The flat surface adjacent to a stream/river composed of unconsolidated fluvial sediment subjected to periodic flooding is very good for groundwater development. Low-lying areas are temporarily waterlogged during floods. This is the youngest geological unit and including various landforms formed by fluvial action. This consists of sand, silt and clays and facilities channel bed infiltration. It is a highly permeable zone helping in partial bank recharge and subsurface flow groundwater occurs under semi-confined to perched water table conditions with shallow water levels. Groundwater prospects in flood plains are almost invariably found to be good (Sharma and Jugran 1992). In this units are exists all along the Cauvery river courses.

Alongside stream channels are relatively flat areas known as floodplain (Fig. 8.10). Floodplains develop when streams over-top their levees spreading discharge and suspended sediments over the land surface during floods. Levees are ridges found along the sides of the stream channel composed of sand or gravel. Levees are approximately one half to four times the channel width in diameter. Upon retreat of the floodwaters, stream velocities are reduced causing the deposition of alluvium. Repeated flood cycles over time can result in the deposition of many successive layers of alluvial material. Floodplain deposits can raise the elevation of the stream bed. This process is called aggradation.

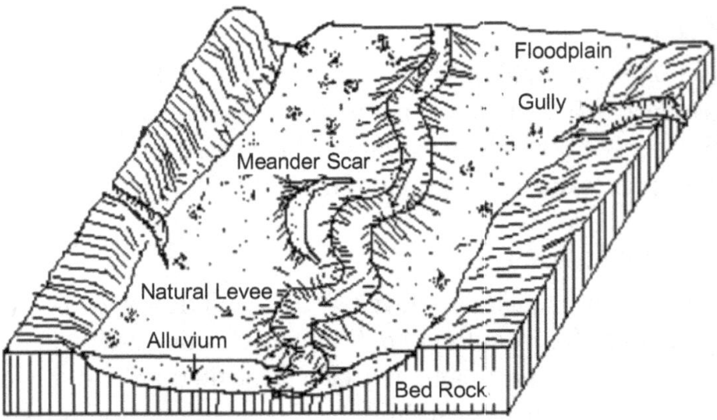

Fig. 8.10: Floodplain and associated features along river channel.

Floodplains can also contain sediments deposited from the lateral migration of the river channel. This process is common in both braided and meandering channels. Braided channels produce horizontal deposits of sand during times of reduced discharge. In meandering streams, channel migration leads to the vertical deposition of point bar deposits.

Both braided and meandering channel deposits are coarser than the materials laid down by flooding.

This unit forms the main source of groundwater in the region. Groundwater can be tapped through shallow and deep tube wells in alluvial plains and flood plains. The wells tapping the flood plains generally give high yield with good quality of water.

Alluvial Plain (Ap)

A flat to gently sloping surface formed by river consist of unconsolidated sediments. It has well to excellent prospect and promising zone or shallow unconfined aquifers.

(i) Younger Alluvial Plains

These are primary sediment storage areas, especially on wide valley floors (Dietrich *et al,* 1982; Nakamura, 1986 and Nakamura and Kikuchi, 1996). This geomorphic unit occurs on either side of the major rivers and their tributaries. The alluvial plain features recorded include point bar (Fig. 3.3 and 8.7) and channel bar deposits (Fig. 3.4). The younger alluvial plain, for example in Purna River basin (Babar 1998 and Babar and Kaplay, 1999a), have an elevation range from 378 to 400 m above msl and characterised by extremely low slope angle (0 to 2) with almost flat surfaces, low drainage density (< 1 km/km^2), low stream frequency (1 to 1.5 streams/km^2) and low relative relief (< 5%).

(ii) Older Alluvial Plain

This unit is seen as irregular patches within the younger floodplains along the river channel. Features associated with this unit are point bar, severe gully erosion in the form of parallel gullies (Fig. 8.11) and meander scars. This morpho unit is located at the foot of the pediments.

The morphometric attributes of older alluvial plain, for example in Purna River basin (Babar and Kaplay, 1999a), are gently sloping (0 to 2), to nearly horizontal surfaces, low drainage density (< 1 km/km^2), low stream frequency (1to 1.5 streams/km^2), low relative relief (*i.e.* between 5% and 10%) and low channel gradient.

The ground water in these units is brackish on account of leaching of salts and calcareous material *i.e.* calcretes are developed into the deeper horizons of these zones. These units as like that of floodplain

Fig. 8.11: Older alluvial plain showing the development of parallel gullies (a) along Dudhana river channel.

landforms form the main source of groundwater. Groundwater can be tapped through shallow and deep tube wells in alluvial plains and flood plains. The wells tapping the flood plains generally give high yield with good quality of water.

Alluvial Fan (Af)

An alluvial fan is a large fan-shaped deposit of sediment on which a braided stream flows over (Fig. 8.12). Alluvial fans develop when streams carrying a heavy load reduce their velocity as they emerge from mountainous terrain to a nearly horizontal plain. The fan is created as braided streams shift across the surface of this feature depositing sediment and adjusting their course. The image below shows several alluvial fans that formed because of a sudden change in elevation.

Deltas (Dl)

Streams flowing into standing water normally create a delta (Fig. 8.13). A delta is body of sediment that contains numerous horizontal and vertical layers. Deltas are created when the sediment load carried by a stream is deposited because of a sudden reduction in stream velocity. Small shifting channels that carry water and sediments away from the main river channel mark the surface of most deltas. These small channels also act to distribute the stream's sediment load

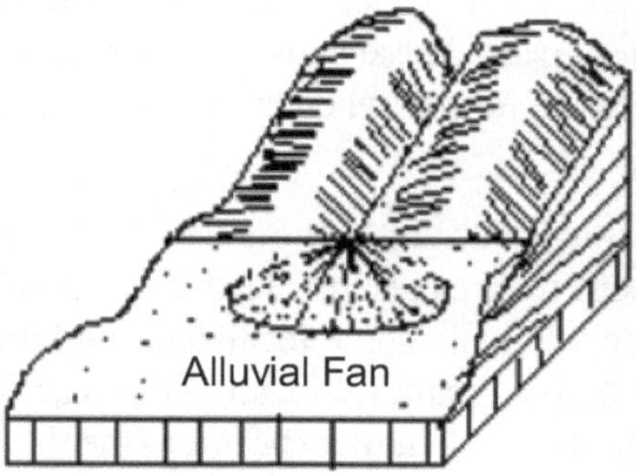

Fig. 8.12: Alluvial Fan

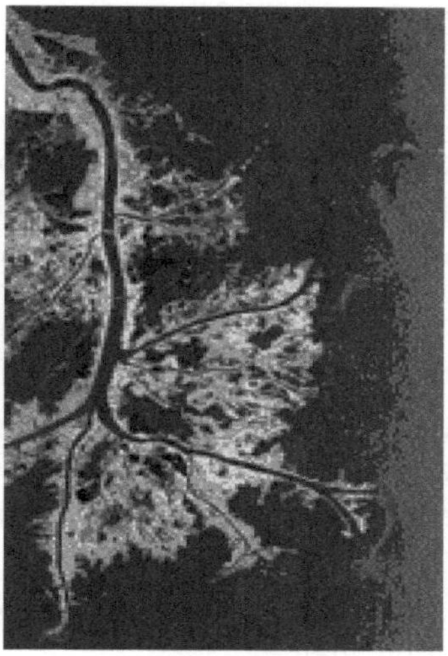

Fig. 8.13: River Delta e.g. Mississippi Delta

over the surface of the delta. Some deltas, like the Nile, have a triangular shape. Streams, like the Mississippi that have high sediment content and empty into relatively calm waters cause the formation of a bird-foot shaped delta.

Most deltas contain three different types of deposits: fore-set, top set and bottom set beds (Fig. 8.14). Fore set beds make up the main body of deltas. They are deposited at the outer edge of the delta at an angle of 5 to 25 degrees. Steeper angles develop in finer sediments. On

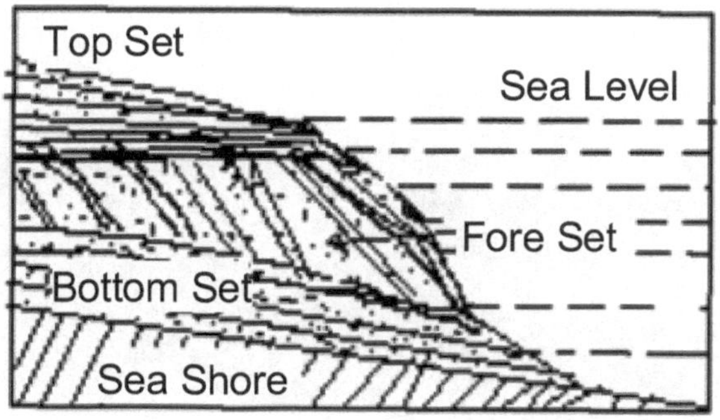

Fig. 8.14: Internal structure of Delta.

top of the fore set beds are the nearly horizontal top set beds. These beds are of varying grain sizes and are formed from deposits of the small shifting channels found on the delta surface. In front and beneath the fore set beds are the bottom set beds. These beds are composed of fine silt and clay. Bottom set beds are formed when the finest material is carried out to sea by stream flow.

Deltaic plain is the major geomorphic unit in the areas of old stage of cycle of river erosion. It is occupy the area comprising of finer loosely packed sediments. Within the deltaic plain some localized low lying areas, affected by waterlogged and soil salinity or alkalinity can be found. The waterlogging and salinity problems may be due to the rising the water level, low lying topography and seepage from canals *etc.*, This zone are found to be south-east, east, north-eastern parts of the study area.

Valley fills (V)

Sediments deposited by stream/ river in a narrow valley mostly fracture controlled. The materials in these surfaces consist mainly of weathered products of the surrounding basaltic rocks; mostly comprise moderately thick gravels pebbles, sand and silt. The prospect varies depending on the thickness of fills. Valley fill deposits are of good potentials for agriculture (Fig. 8.15).

The morphometric attributes of the valley fill deposits (Babar and Kaplay, 1999a) found in Purna river basin are: moderately steep gradient

Fig. 8.15: Valley fill deposits at Sawangi (Jintur) Maharashtra.

(5^0 to 10^0), moderately coarse drainage density (2 to 2.5 km/km^2), moderate stream frequency 2.5 to 3 streams/km^2) and moderate relative relief (20%). Valley fills act as good groundwater potential zones and water table is shallow. Valley fills has very good groundwater prospects as it is being recharged by the surrounding hills as well as by the river water. Deep bore wells are preferable in these zones for intensive agriculture.

River Sand

Sandy material deposited along stream channel act as very good recharge zone for groundwater.

Ravines (RA)

Small narrow, depressions usually carved out by running water. They have poor groundwater yield.

Meander Scar

These are the remnants of the highly sinuous paleo-drainage system, which were cut-off from the main channel. They can be identified in the satellite imagery by arcuate shape, uniform tone, isolated occurrence and depressed relief characterised by the loss of hydraulic continuity with parent channel. A crescent shaped scars of meandering stream (Fig. 8.10) still discernible on surface has excellent potential for ground water.

Old Meander

Abandoned meandering loop of river/streams has excellent potential for ground water.

Paleo-channels

These are the ancient drainage lines of streams or rivers through which it might have flown in past. Paleo-channels can be recognised on the satellite imagery by their medium to darker tone, curvilinear pattern, uniform texture and continuity of older stream. The remnant of stream/river appears, as buried or abandoned channels are promising zone for shallow aquifer with excellent ground water yield and hydro-chemically the water is potable.

Back Swamp

A low-lying swampy/marshy area adjoining natural levee has good prospect.

Pond/Lake

Ground water recharge also takes place through tanks/ponds/ reservoirs and check dams besides the natural mechanism all maps show this feature.

Landforms of the Coastline (Marine).

Sea Beaches

Abundant broad deposits of sand on a shoreline form beaches. These may be cut shoreward during stormy activity, but are replaced during long periods when wave action is weak. As a wave breaks, a powerful surge of turbulent water runs up the beach called swash. The return flow is called the backwash. This strong underwater current is sometimes called an undertow and can be strong enough to carry people away from the beach. Waves, however, rarely attack a beach at a perfect right angle, but rather approach a beach at less than a right angle. Hence, a breaker will push sand up at an angle, but the sand will move down the beach at or greater than a right angle. This causes the beach sand to slowly move up-beach, and the process is called *beach drift*.

The sediment in the shore zone is also transported up-beach in a process known as *longshore* drift by the longshore current which travels at a right angle to the beach. Together, these processes are called *littoral drift*, where "littoral" refers to the coast or shoreline. Over long periods of time, mineral particles or sediment can be transported great distances.

Where a bay exists, the transported sand will collect at one end and form sand spit. Where sediment is eroded from a headland and deposited inside the bay, a pocket beach is formed. The process of active beach building occurs when more sand arrives at a beach than is taken away, a process called progradation. When more beach sand is taken away than arrives, the process is called retrogradation.

To build an artificial beach, a groin can be installed. These are walls or embankments, which catch the beach drift to form a beach.

Barrier-island coasts

Along the low-lying coastal plains of the southeastern U.S. lie barrier-island coasts, formed from low ridges of sand built by waves and further increased in height by the growth of sand dunes. These act as barriers, preventing severe storms from causing more extensive damage inland.

Lagoons

Barrier islands have lagoons behind them, broad expanses of shallow water filled with tidal deposits. Gaps between barrier islands are called tidal inlets, which allow the tidal current to reach the inland shore.

Coral reef coast

Corals secrete a carbonate mineral, which build up to form a coral reef coast, a unique coast in that living organisms form it. These occur in warm subtropical waters between 30°N and 25°S.

Atolls

Associated with coral reefs are atolls, circular coral reefs that enclose an interior body of oceanic water. These form when a pre-existing volcano is eroded down, leaving behind only the outer coral reefs.

Marine terrace

Sometimes a shoreline can be suddenly raised as in an earthquake, resulting in a raised shoreline, with a resulting marine terrace.

Denudational Landforms

Four agents exist on the earth, which carry out the process of denudation: (1) water, (2) wind, (3) ice, and (4) waves. Denudation is the total action of all processes by which exposed rock is worn away and the resulting sediments are transported to the sea or closed inland basins. Denudation therefore acts to lower all continental surfaces to a flat featureless landscape. However, tectonic forces are constantly building new landforms for denudation processes to wear down.

Pediment (P)

A Gently slopping smooth surface of erosional bedrock with thin veneer of detritus is called pediment. Ground water prospects are moderate to poor at the intersections of lineaments.

Pediment is isolated residual hillocks being remnants of weathering and denudation. Inselbergs are mostly barren, rocky, usually smooth and rounded small hills. From groundwater point of these are all treated neither containing nor transmitting of waters *i.e.* aquifuge nature. Mostly acts as run off zones.

This geomorphic unit is basically pediment zone and having various soil types cover over the pediment materials. This unit is characterized by very high runoff zone and poor groundwater recharge horizons.

In this geomorphic unit there is the gently sloping and undulating rock surface with or without a veneer of weathered material. Pediments are noted as a narrow strip adjoining the highly dissected plateau (Fig. 8.16) and at foothill zones. (Babar and Kaplay 1999a) found the pediment zone with range in elevation from 400 to 450m above msl and morphometric attributes of the pediments are gently sloping surfaces (2^0 to 5^0), low gradient, moderate drainage density (2 to 2.5 km/km^2)

Fig. 8.16: Pediment surface along highly dissected plateau in Jintur, Parbhani district, Maharashtra.

moderate stream frequency (2.5 to 3 streams/km^2) and relative relief is about 15 to 20%. Groundwater potential in pediments is poor except along fractures where limited quantity of groundwater can be obtained for domestic purpose.

Pediplain

This geomorphic unit is developed as a result of continuous processes of pediplaination. The altitudinal variations is relatively high for rolling plain and is about 5 -10 m. In this horizon there exists irregular dissected portions with a number of gully are present. The pediplain with sedimentary rocks exposures are generally formed due to intensive weathering under semi arid climatic conditions, representing final stage of the cyclic erosion (Knig, 1950 and Sparks, 1960). These are identified in the imageries with grey tone on false colour composite (Ghose 1993). Groundwater prospect in this unit is good due to the moderate thickness (15-20 m) weathering materials (Babar and Kaplay 1999a and Shah 2002).

Pediplains are undulating to gently sloping part of the basin with slope angles of 0^0 to 5^0 and lies parallel to the stream course. It has low drainage density (1.5 to 2 km/km^2) and elevation range of 400 to 420 m above msl. This geomorphic unit is potential for intensive agriculture. The soils of this zone are fairly thick, dark grayish brown to black in colour, calcareous in composition, moderately well to ill drained and high in moisture retentive capacity and are placed in 7th Approximation of soil classification under class "Vertisol" (Malewar, 1976).

Pediplain are found to be good for groundwater potentiality. The weathered zone thickness range from 10 to 15m. The irrigation in this zone can be done mainly through dug wells.

Various types of buried pediplains are described below:

Shallow weathered Buried pediplain (BPP-S)

A flat and smooth surface of buried pediplain with thickness of 0 to 5 m consisting shallow overburden of weathered derivative material. Ground water prospects are moderate to poor but open wells yield good amount of potable water after monsoon.

Medium weathered Buried Pediplain (BPP-M)

A flat and smooth surface of Buried-Pediplain with moderately thick 5 to 20 m overburden of weathered derivative material and has good potential for groundwater and less seasonal variations in water table is observed. This units is moderately weathered with a thickness varies from 5-20 m. The groundwater prospects in these units are moderately good. In this zone are found close of the Cauvery River, south central, southwest and southern parts of the study areas.

Deep weathered Buried Pediplain (BPP-D)

A flat and smooth surface of buried pediplain with very thick i.e. more than 20 m overburden of weathered derivative material and has good potential for groundwater and less seasonal variations in water table is observed. In this unit infiltration is moderately good. The thickness of weathered zone varies from 10-20m and favours a good amount of water to circulate within this zone before reaching the deeper fracture zone. Groundwater potential zone is very good and this unit is suitable for dug-well, dug-cum-bore wells and bore wells.

Denudational hills (DH) and residual hills (RH)

Denudational hills are marked by sharp to blunt crest lines with rugged tops indicating that the surface run off at the upper reaches of the hills has caused rill erosion. They can be interpreted from their massive size and glottal to elliptical shape. The rugged topography of this region is due to the erosion of the denudational hills to the plain region, leaving the rock exposed. The groundwater potential is moderate to poor.

Residual hills are the end products of the process of pediplanation, which reduces the original mountain masses into a series of scattered knolls standing on the pediplains (Thornbury 1990). The more resistant hills stand out prominently under the condition of differential erosion and weathering. It's occurring as isolated patches are found at lower altitudes complex to the denudational hills. In spite of their isolated occurrence, their continuity in a linear or curvilinear fashion gives indication that they are structurally controlled. The shapes of the residual hills are controlled by the different factors such as lithological composition, distribution and spacing of joints and fractures. In the imageries dark grey tone and coarse texture in block and white images and dark reddish colour in false colour composite with radial drainages pattern (Gupta 1980). Groundwater prospects are believed to be poor in the areas. It serves as a good recharge zone but cannot hold groundwater due to steep hydraulic gradient.

A group of massive hills with resistant rock bodies that are formed due to different erosional and weathering processes are called Denudational hills. The valleys are seen in between these hills, which are, filled with boulders, sand and silt. These hills have the steep gradient, high drainage density, high stream frequency, higher relative relief and are characterised by rapid runoff.

The soil cover over this geomorphic unit is very thin. The soils are rocky, clayey and underlained by weathered basalts. The hills are sometimes barren and at places clothed with the forest. The yield of groundwater in this area is very poor.

Structural Landforms

Escarpments

Escarpments are the structurally controlled erosional features produced by streams in regions composed of alternating beds of hard and soft rocks. The stream easily and quickly erodes the soft layers of rocks whereas the hard layers resist the erosion and stand projecting as ledges on the sides. These ledges are under cut by continued erosion by the stream water. A time comes when a given ledge is not able to support itself any further and falls down into the river giving rise to steep slope. It is the slope caused by falling undercut ledges of hard rocks that is referred as an escarpment (Fig. 8. 17). A steep vertical cliff separates alluvial and hard rock terrain has poor groundwater prospects.

Dip Slope

In some cases the stream may succeed in completely eroding the overlying soft layer, there by exposing the under lying hard rock all along its dip. The resulting slope is then same as the dip of dip of the layer and is in itself a typical erosional feature. It is called a dip slope (Fig. 8.18).

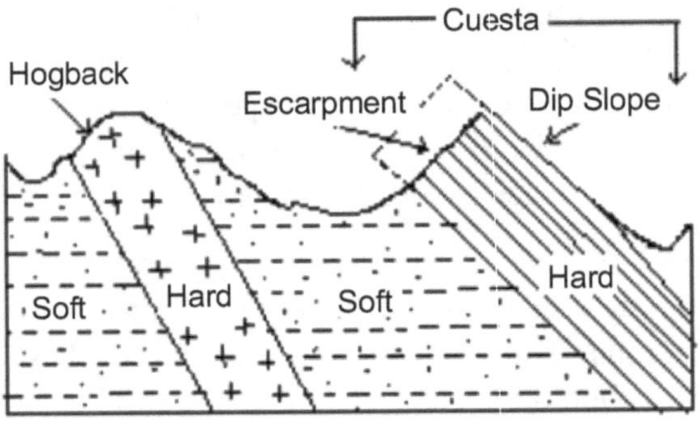

Fig.8.17: Escarpment and its related features.

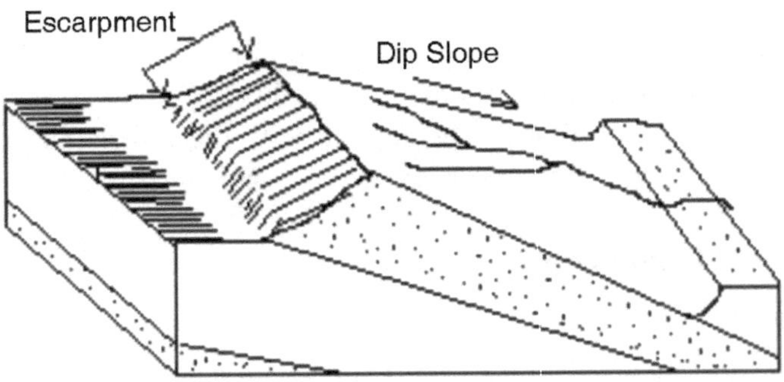

Fig. 8.18: Development of escar pment and dip slope.

Cuesta

It is the term given to a combined set of escarpment and dip slope occurring adjacently. Obviously, it result due to prolonged erosion of rocks forming the channel of the river and having an alternate hard and soft rock layers (Fig. 8.17). This unit can be identified on satellite imagery by its darker tone, linear pattern, isolated hillocks, parallel pattern of first order channels and scanty or no vegetation. It exhibits asymmetrical ridge profile and is controlled by gently dipping strata. In this unit groundwater prospects is poor.

Hogback

It is an erosional feature developed by river action. It is essentially an outcrop of a hard resistant rock that is very steeply inclined (Fig. 8.17). In satellite imagery it displays the tonal banding, linear forms, parallel drainage and symmetrical profile. It has erosional slope on either side. It shows scanty vegetation cover and high resistance to erosion. It has poor groundwater prospects.

Structural Hills

Structural hills are the linear or arcuate hills exhibiting definite trend lines. These hills are structurally controlled with complex folding, faulting, criss-crossed by numerous joints/fractures which facilitate some infiltration and mostly act as run off zones. These are found to be north to northwestern parts of the study areas. The structural trend of the hills ranges in the southwest to northeastern direction with slight deviation towards the western part. The slope of the hills ranges from 2 degree to 4 degree. The structural hills have poor to nil ground water prospects because most of rainwater, which falls over them, goes down-slope as surface run-off.

Chapter 9

Applications of Hydrogeomorphic Studies

⇒ The applications of hydrogeomorphic studies are discussed with reference to the groundwater potential zones, groundwater management, water resources assessment, watershed development, artificial recharge studies and hydrogeomorphic assessment method for wetlands.

Water is a primary source of life and sustains all human activities such as domestic needs, agriculture, industries *etc*. The allocation and management of water resources is becoming a difficult task due to increasing demands, decreasing supply and diminishing quality. This calls for judicious use of water resources. The geomorphology describes the environment in which the hydrological processes operate. A strong mutual correlation exists between geomorphological variables and hydrological characteristics. Such relationship can be applied to both surface and groundwater regime. Thus the linking of geomorphological parameters with hydrological characteristics of the basin provides a simple way to understand the hydrogeomorphological behaviour of different basins and particularly of un-gauged basins.

During the last few years various applications of hydrogeomorphology with reference to the Geographic Information System (GIS) and various other techniques have emerged as an extensively effective tool for analyzing and prioritizing natural resource management alternatives. Because natural resource management problems are spatial in nature the GIS technology provides a tool for defining the extent of the problem, and facilitates the design and implementation of alternative management strategies. The GIS is used to estimate the hydrogeomorphological parameters of the watershed, which can be used in watershed planning and management at micro and macro levels. The results of the study can be utilized to infer the hydrological features like runoff potential, infiltration and groundwater recharge of the region. Some of the applications of hydrogeomorphological studies are described below:

Groundwater potential zones

The most important application of the hydrogeomorphological studies is the assessment of groundwater potential zones of the watershed or a basin. At first groundwater potential feasibility map of the area is to be prepared using multi-criteria analysis. Multi-criteria evaluation is primarily concerned with how to combine information from several criteria to form single index evaluation. The combination of various techniques and the data sets would be utilized for targeting feasible potential zones of groundwater. The data sets include, geology map, lineament map, slope map and soil map. Above all the hydrogeomorphological map of the area must be prepared to arrive at the firm conclusion. On the basis of relative importance, a data set has been chosen for different information layers and best suitable condition is derived.

In the soft rocks and in the loose unconsolidated rocks the occurrence of groundwater is controlled by the primary porosity in the form of voids and pore spaces. In the hard rock terrain, secondary features like fractures and weathered rock control the occurrence of groundwater. From the feasible groundwater potential map it is seen that the most potential areas for groundwater matches with the pegmatite and dyke, which are considered to have good aquifer storage.

The population growth has been creating more and more stress on agriculture sector for increasing the food grain production, which consequently increased deforestation and demand for more water. The available surface water resources are inadequate to meet the entire water requirement for various purposes. So the demand for underground water has increased over the years. Generally, groundwater is less prone to pollution in comparison to surface water. Hence, groundwater serves as an important source of water for various purposes in rural and urban areas. In recent years intensive use of satellite remote sensing has made it easier to define the spatial distribution of different groundwater potential zones on the basis of geomorphology and other associated features (Sinha *et al,* 1990; Kulkarni 1994; Adyalkar *et al,* 1996; Goswami *et al,* 1996; Srinivasa Rao *et al,* 1997; Bhan 1998; Bhagwan 1998; Bhagwan and Raghu, 1998, Patil *et al,* 1999, Babar 2001 and Babar 2002a and b).

In many earlier studies (Lattman and Parizek 1964; Bedi and Bhan 1978; Karanath and Seshu Babu 1978; Raju *et al,* 1985; Satyanarayana 1991; Palanivel *et al,* 1996 and Srinivasa Rao *et al,* 1997) remote sensing techniques have been applied for estimation of groundwater potential zones.

Babar, (2001, 2002a and b) while studying the hydrogeomorphology, of the Akoli watershed (Jintur tehsil, Parbhani district, Maharashtra) has divided into areas occupied by valley fill, pediplains, pediments, highly dissected plateau and denudational hills. By studying the hydrogeomorphological conditions of the basin, it is possible to decipher the groundwater potentiality. It is moderate to good, in most part of the watershed while the fractured zones in the weathered pediplains and valley fills are very good potential zones (Babar 2002a and b). The pediplains and pediments with fractures are having moderate groundwater potentiality. Dug wells are recommended in the weathered pediplains. Most of the dug wells in the pediments and in pediplain dry up during pre monsoon period hence bore wells and dug-cum bore wells are recommended for these areas.

Groundwater management

For optimal utilization and management of the limited groundwater reserves, to meet increasing demand for supplementary irrigation and domestic need and also for accelerating the existing recharge phenomenon in the watershed, detailed information for basement topography, aquifer geometry and fracture systems are essential. The various approach and techniques (as discussed in Chapter 10) can be adopted for deriving above information; similarly another very important technique is the Digital Basement Topography Modeling (DBTM) as suggested by (Kumar and Prasad 1993, Kumar *et al,* 1996 and Kumar and Prasad, 2003). This has helped in understanding groundwater storage and retrieval system for sustainable utilization of groundwater in close relation with surface water. The over all objective of the study was to provide scientific database for sustainable utilization and development of groundwater and surface water. Estimation of groundwater reserve of entire aquifer system is essential for sustainable planning of utilization and development of groundwater. Present planning process only looks into utilization and development of replenished groundwater. Utilization of groundwater, beyond the limit of replenishment, will create imbalance in intake and outtake to the aquifer and it will result into depletion of water table. The increased utilization can be supplemented from utilizable groundwater available in aquifer between dug-well base and basement surface. For this, proper mechanism has to be developed to increase the recharge to the aquifer so that the balance between intake and outtake to the aquifer can be maintained.

At present only 5-10% of annual rainfall is contributing to the groundwater through natural recharge process. Further, due to unconfined aquifer system, nearly 40-60 % of the total annual natural recharge gets lost through nalas/drainage channel. Therefore there is need to utilize the available reserves before it get lost as seepage or check the seepage loss through appropriate method. To meet the increased demand beyond the replenished limit, groundwater recharge is to be increased by accelerating the existing recharge process.

Water Resources Assessment

Water resources assessment of a region involves a detailed study of the surface and sub-surface water. To integrate the entire surface and sub-surface data manually requires huge manpower and time. By adopting a GIS platform and other techniques the result obtained will be faster and more accurate. Till recently, ground water assessment was based on laboratory investigation, but the advent of Satellite Technology and GIS has made it very easy to integrate various databases.

Groundwater is a most important natural resource required for drinking, irrigation and industrialization. The resource can be optimally used and sustained only when quantity and quality of groundwater is assessed. It has been observed that lack of standardization of methodology in estimating the groundwater and improper tools for handling the same, leads to miscalculation of estimation of groundwater. It is essential to maintain a proper balance between the groundwater quantity and its exploitation. Otherwise it leads to large-scale decline of groundwater levels, which ultimately cause a serious problem for sustainable agricultural production. A possible solution for such problems is micro level planning, and use of standard hydrogeomorphological methodology for assessing the groundwater. In recent years micro level planning has gained acceptance, since it can be locally applied and readily managed by self-sufficient rural governance.

In the current development, Ground water assessment is based on the water table fluctuation method and calculated based on the norms recommended by National Groundwater Estimation Committee (NGEC). The parameters used include:

- Normal Monsoon Rainfall data for 20 years
- Normal Yearly Rainfall data for 20 years
- Depth to Water table for Pre and Post monsoon water level
- Specific yield data for the zone of water table (this is expressed as fraction and varies with Geological formation)
- Geomorphology Map
- Canal Map
- Surface Water Irrigation (For Monsoon and Non-Monsoon)
- Return Seepage from Ground Water Irrigation during Monsoon
- Surface Reservoir map / Water body map and other conservation structure
- Well Data
- Landuse/land cover map
- Slope map

Using above all parameters the hydrogeomorphological map must be prepared so as give the correct information of the groundwater resources. By using these parameters the easier assessment of groundwater resources is possible (Kulkarni H 1994; Adyalkar P G *et al,* 1996; Goswami D C *et al,* 1996; Srinivasa Rao Y *et al,* 1997; Bhan S K 1998; Bhagwan S V B K 1998; and Patil B S *et al,* 1999).

Watershed development

Natural resources development and management on the basis of watershed at micro level has been adopted by most of the agencies for sustainable rural development. A Watershed is a hydrogeomorphological unit, in which the streams are draining to a common point. The physical parameters, which influence the nature and characters of watershed, are geology, geomorphology and hydrogeology. All these factors influence the movement and distribution of water (surface and ground).

The population outburst and rapid technological advancement have put excessive pressure on the available water resources. This has led to water scarcity and which eventually reduced crop productivity. Accurate knowledge and description of physical framework is essential for monitoring of the watershed. A proper understanding of the problem and simple cost effective method is required to reduce the impact of water scarcity and its socio-economic aspects.

The Concept of watershed as a planning unit for development of land and water resources has gained importance since 1974, when the Ministry of Agriculture, Government of India initiated various development programs like Drought prone Area Programme (DPAP), Desert Development Programme (DDP), Hill Area Development Programme (HADP) *etc.* Therefore, it is necessary to delineate watershed boundaries at various levels of hierarchy to identify development activities under various schemes in each watershed.

Watersheds are natural hydrogeomorphological entities that cover a specific areal extent of land from which rainwater flows to a defined gully, stream or river at any particular point. The size of the watershed is dependent on the size of the stream or river, the point of interception of the stream or river and the drainage density and its distribution. The All India Soil and Land Use Surveys (AIS and LUS) of the Ministry of Agriculture, Government of India have developed a hierarchical system of watershed delineation like water resource region, basin, catchment, sub-catchment and watershed. However, for the study IMSD (1995) has decided the following six levels of watershed delineation:

1. Catchment
2. Sub-Catchment
3. Watershed
4. Sub-watershed (area 30-50 km2)
5. Mini-watershed (10-30 km2)
6. Micro-watershed (5-10 km2)

In view of the limitations of the total water resources availability due to peculiar agro-climatic and hydrogeological conditions prevailing in the area it is essential to plan out a systematic and scientific programme of development of water resource through an integrated approach of water harvesting conservation and augmentation by artificial recharge as well as by conjunctive use of surface and groundwater resources. The water resources management and conservation should be planned on micro-watershed basis laying emphasis on capturing the water in-site for its optimum use within the watershed itself.

The natural resources like land, soil and water are to be used optimally in a planned self-sustaining manner so that the natural environmental balance is not adversely affected. This necessitates systematic micro-level delineation of different hydrogeomorphological parameters. Fore most amongst these resources is water. The monsoon runoff in the area invariably goes waste. Construction of stream barriers and filtration wells will help checking this and recharging the groundwater. These artificially developed percolation wells and pits can conserve great deal of water. The construction of underground bandhara (detail of which is given in next section of this chapter) is necessary to arrest subsurface flow from watershed to protect and conceive base flow runoff from superficial zones. The increased storage of groundwater thus created is expected to cater domestic, agricultural and industrial water requirements of the area and promote multidimensional human activities. The additional benefits of water conservation would be minimal soil erosion, flood control and conservation of ecology of the region.

Repeated water scarcities leading to large-scale droughts have severely affected the livelihoods of the rural poor in India. Three types of reactions to such situations can be broadly observed:

(a) Short term, relief measures to mitigate water shortages by developing water sources that are often not sustainable.

(b) Highly expensive measures involving relief measures like supplying water through tankers (as relief measures) and development of regional piped water supply schemes that require high costs.

(c) Local solutions through participatory approaches that have increasingly resorted to the integrated watershed management model to identify, assess and address the larger problem of rural systems management.

Natural resource mobilization and management along with watershed lines is becoming a key strategy for rural development and poverty alleviation in rain-fed areas. The emphasis in watershed development is to enhance the productive base of rain-fed agricultural and biomass based livelihood systems. This can be achieved by arresting soil degradation and improving groundwater availability by recharging aquifers through in-situ rainwater harvesting and improving the ecology by plantation of trees, grasses and natural regeneration.

Watershed development as a comprehensive and inter-sectoral activity, with the active participation of primary stakeholders requires capacity building and handholding in technical, social, gender and managerial aspects. This is required not only during the watershed development project implementation, but also for creating necessary competencies and skills for maintenance of assets created. It also involves setting up and sustaining of vibrant and empowered local institutions and self-help groups. Timely credit flow, without cumbersome procedures are also required for better utilization of developed resources, be it for improved agricultural activities or allied income generation activities.

Since the watershed developmental programme are implemented in a spatial domain, it is necessary to have a sound resource database/ thematic maps on geology, hydrogeomorphology, soils, landuse / land cover and land degradation status etc. This could be achieved through Remote Sensing technique, which provides data in different combinations of spatial, spectral and temporal resolution. Remote Sensing being multi-disciplinary, single satellite imagery provides information on many aspects of the watershed. The spatial database/ thematic maps pertaining to soils and land capability, landuse/land cover, hydrogeology, groundwater potential (hydrogeomorphology), slope, drainage, watershed boundaries, transport network, village boundaries would be immensely useful as baseline information for watershed management. These thematic maps provide detailed information regarding major characteristics of watershed. Most of the natural resources database is being generated using satellite data. Hence, the satellite imagery will act as baseline information during the impact assessment studies using Remote Sensing techniques.

The pressure on the country's natural resources (land and water) has been continuously and steadily increasing mainly due to increase in population and people's aspiration for improved quality of life (Asthana and Asthana, 2003). Soil erosion and deforestation are the other major factors limiting the land productivity. Out of 329-mha geographical area of the country, about 175 mha suffers from one or the other kind of land degradation problem. On the other hand, out of the average total annual rainfall of 400 million hectare meters (Mham) only 60 Mham is used while the rest goes as run off. In addition, due to uncertain availability of water for use and frequent droughts in different parts of the country, the productivity becomes unpredictable. Hence there is a need of perfect well-planned watershed development and management schemes adopted with reference to the hydrogeomorphological aspects.

Artificial Recharge Studies

With the increasing use of groundwater for various activities the fast decline in groundwater takes place. In order to prevent the aquifer from fast depletion, the artificial recharge is very much essential. Keeping this in mind to avoid fast decline in groundwater levels in various parts of India, the studies are being undertaken to identify the favourable areas for artificial recharge and suggest suitable recharge structures to augment the aquifer system (Athavale *et al,* 1992, Bhatt and Salpekar, 1990, Karanth 1989, Mahajan 1993, Murlidharan and Athavale, 1998, Raju, 1985, Raju, 1998, Ramanuju Chary and Subbarao, 2002, Ratnaparkhi 1975, Sharma *et al,* 2000 and Thorat and Sable 1990). The analysis were carried out block wise for the different states Andhra

Pradesh, Karnataka, Madhya Pradesh, Maharashtra and Tamil Nadu using Remote Sensing data and GIS techniques. The various thematic maps such as geomorphology, geology, soil, slope, landuse, drainage, drainage density, lineament density, runoff isolines, depth to weathered zone, depth to basement, groundwater level fluctuations and water quality were used in the analysis. The above maps were prepared using satellite data and other collateral information collected from the field and digitized data available.

Considering the terrain conditions and favourable zonation, the suitable artificial recharge structures such as percolation pond, check dam, recharge pit, recharge shaft, contour trench, nala bund, recharge tube wells and subsurface dyke can be recommended. These recommendations must be communicated to district administration for implementation. The techniques suggested in this chapter are to arrest the surface runoff and to add additional water to groundwater.

The integrated approach as suggested will help in strong the surface runoff to replenish the underground storage. The rainwater harvesting can be planned in a small/micro-watershed comprising an area of 10 to 20 sq. km. in a group of 3-4 villages. (Sara Naaz and Venkateshwara Rao, 1998). The watershed can be selected by remote sensing imagery and after detailed ground truth mapping, following structures can be made to replenish the subsurface storage:

1. Continuous contour trenching
2. Percolation tanks
3. Series of weir across streambed by natural stream channel method.
4. Nala bunding
5. Underground bandhara
6. Recharge trench.

The salient features of these structures are discussed below:

(i) Continuous contour Trenching (CCT)

This can be undertaken along the foothills, hilltops and divide lines of watershed along the Contour. The water can be arrested in the trenches during the rainfall. The stored water will slowly infiltrate in the down stream area and thus groundwater storage can be augmented.

Along the terraces in these areas the plantation is proposed after making artificial cracks by bore hole blast technique so that the rate of survival of the plants would be more. The method of CCT is extensively practiced in the area around Yeldari reservoir in Jintur tahsil (Fig. 9.1). Such extensive CCT are suggested in the watersheds for rainwater harvesting.

Fig. 9.1: Highly dissected plateau showing continuous contour trenches

(ii) Percolation Tank

Depending upon the natural catchment available and artificial catchment, which can be maple available, the capacity of percolation tank can be decided and 2 or 3 percolation tanks can be constructed in the upstream side of surface drainage in runoff zone (Sardhukan, 1990). The percolation tank will hold immediate runoff at the time of rainfall and will accommodate huge quantity of surface water, which could have other wise gone as surface runoff.

(iii) Series of weir Across Streambed by Natural stream channel method

In this method the artificial recharge is made to restrict the surface runoff through nala or stream by making additional water available for percolation by constructing a weir across the bed where the stream has a wide bottom of the channel. The surface water impounded during monsoon behind the weir and spilling over it spreads out in shallow depth over the entire streambed and thereby increasing this wetted area. The impounded water increased the water level of base flow in the stream. So the water level of the groundwater increases and it can be stored for use in dry season. A series of weirs can be constructed in the streambeds (Fig.9.2) such that over flow of the upstream across weir can fill the downstream area. Such series of weirs are fruitful in the area where the weathered mantle should be of 3 to 6 m thickness.

Fig. 9.2: Photograph showing A series of weirs constructed in the streambeds at Shevdi, Jintur, Maharashtra.

(iv) Nala Bunding

To retain the flow stream, nala bunding (as like weir) is constructed (Fig. 9.3). Part of stored water goes as a recharge to subsurface. The series of such nala bunding if made at suitable sites can give good amount of storage for Rabi (winter season) cropping.

Fig. 9.3: Check dam as nala bunding measure at Jintur, Maharashtra.

(v) Underground Bandhara

If the area consist of thick sediments, soils or weathered (>6m) material, the site can be suitable to locate or construct the underground bandhara (Fig. 9.4). Such suitable sites if available for arresting subsurface runoff, the sites are selected in the watershed area and through underground bandhara of impervious clay the movement of groundwater below the bed level stream are arrested to make a sub-surface storage. Fig.9.5 illustrate the internal structure of the underground bandhara through cross section.

OVERVIEW OF UNDERGROUND BANDHARA

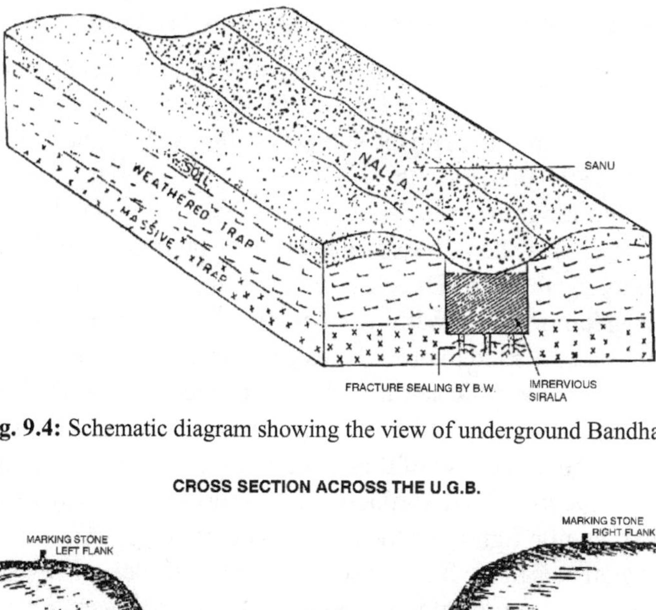

Fig. 9.4: Schematic diagram showing the view of underground Bandhara.

CROSS SECTION ACROSS THE U.G.B.

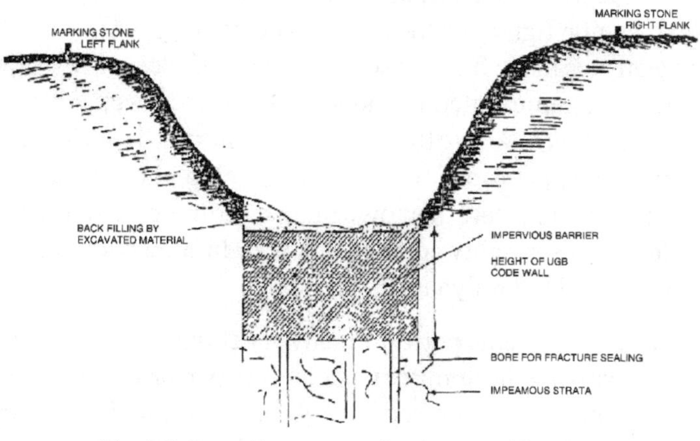

Fig. 9.5: Internal structure of underground bandhara.

(vi) Recharge Trench

Recharge trenches are to be constructed across the slope in contact with weathered or hard rock to a depth of hard dense basement. The design of trench can be framed depending upon the suitability of structural and hydrogeological control. During the rainy season the runoff of rainwater can be arrested and allowed to accumulate in the trench. The arrested water can augment the groundwater from downstream area and thus will help in building more storage. The suitability of the recharge trench will be more in over – exploited areas where the depth of weathering is more than 10m in hard rock area.

Hydrogeomorphic Assessment Method For Wetland

The growing interest in the hydrogeomorphic assessment method for wetland functions is reflected in the increasing number of publications and research on the subject (Brinson 1993 and 1996, Babb 1997, Cole 1997 and Huberth 1994). Wetlands can be classified by methods that range from the use of commonly recognized vegetation or cover types, to systems based on hydrology, geomorphology, or some combination of the two. The classification presented here is based on the hydrogeomorphic functions of wetlands. There are three basic properties that are used to provide insight into wetland functions. With reference to geomorphic setting - the three categories of wetlands are depressional, riverine, and fringe. Extensive wetlands constitute a separate category because of their unique topographic and hydrologic conditions. Depressional wetlands can be open or closed to surface flows, and can be tightly or loosely connected to groundwater flows. Riverine wetlands range from those associated with steep to low gradient streams and are represented by floodplains. Fringe wetlands are sea level or lake level controlled. Wetlands normally initiate their development in depressions. If wetlands develop beyond the original depression, they can create their own unique geomorphic settings. Each of these four types roughly corresponds with limited combinations of water sources and hydrodynamic conditions.

Wetland ecosystems perform many functions that are beneficial to society. Assessing the ability of wetlands to perform these functions is an important part of the regulatory, planning, and management activities of various implementing agencies. Various methods have been developed for assessing the functions of wetlands; however, none has received widespread acceptance or use within the Corps because of a basic failure to meet both the technical and programmatic requirements. As a result, the assessment of wetland functions is often completed using subjective and, in some cases, undocumented methods. The procedure to assess the functions of wetlands begins with the classification of wetlands into "functional classes" based on hydrogeomorphic characteristics. For each wetland class, a suite of appropriate functions is identified along with functional indicators. Functional indicators are used to characterize the physical, chemical and biological properties and processes of the wetland and the surrounding area and to predict the functional capability of the wetland in the context of a reference wetland population.

Classification of wetlands is the assignment of wetlands to groups within a system of categories distinguished by origin, structure, flooding frequency, dominant organisms, or some other combination of physical and/or biological attributes. Hierarchical classification of wetlands is a system of classification where wetlands are ranked in categories one above another (Brinson 1993, Huberth 1994 and Smith 1995). This classification can be modified and expanded formally to include water regimes, water chemistry, hydrogeomorphic units and substrate as dominant type categories. Brinson (1993), Huberth (1994) and Smith, (1995) provided modifiers for the classification in the form of categories for water regimes, chemistry and dominance types. Thus wetland can be classified based on its ecosystem context such as lagoons, pools, ponds, seeps, springs, etc. It is necessary to have a classification of wetlands that can adequately portray the richness of wetland types. This is essential if conservation efforts are to preserve at least a representation of the wealth of the areas natural heritage.

Chapter 10

Techniques in Hydrogeomorphology

⇒ In this chapter various techniques in hydrogeomorphology are discussed including conventional techniques, remote sensing, Aerial photo-interpretation, Geographical Informational system (GIS), Global Positioning System (GPS), Digital Elevation Model (DEM), Groundwater Information System (GWIS), Digital Surface Terrain Modelling (DSTM), Multi-criteria evaluation (MCE), Thermal Infra Red (TIR) Mapping and Microwave Data and Side-Looking Airborne Radar (SLAR).

New techniques and approaches to groundwater studies are always in deep concern with the hydrogeomorphologists. The designing of new techniques or adapting existing techniques with better enhancement to solve practical problems in the study of groundwater are of central concern. One of the recent technical developments include the mathematical model of groundwater flow that has become the most widely used computer-based model in the groundwater industry. Techniques to identify and date recently recharged water that is useful in characterizing the susceptibility of hydrogeologic environments to contamination are utilized by various research groups all over the world.

Geophysical methods for determining the hydraulic properties of fractured rocks gained the importance in the hydrogeomorphic studies. Predicting the movement of water and contaminants through fractured rocks has been one of the most challenging problems in hydrogeomorphology. Besides conventional and above mentioned techniques the techniques discussed in this chapter are Remote Sensing, Geographical Informational system (GIS), Global Positioning System (GPS), Digital Elevation Model (DEM), Groundwater Information System (GWIS), Digital Surface Terrain Modelling (DSTM), Multi-criteria evaluation (MCE) and Thermal Infra Red (TIR) Mapping and Microwave Data.

Conventional techniques

Groundwater is a fairly widely distributed natural, renewable resource, which gets replenished almost regularly by precipitation. This renewable source needs to be managed in such a way that the withdrawal is kept at pace with the recharge. Groundwater occurrence at any place is a consequence of interaction of climatic, geologic, hydrologic, physiographic and ecological factor. The search or groundwater is mostly confined to promising zones in terms of porosity and permeability of the rock formations.

The conventional methods of groundwater prospecting and the hydrogeomorphological surveys include well inventory, mapping of aquifer characters, water sample analysis etc. These methods are suitable for evaluation of shallow aquifers, which have been partly or fully developed through, open wells. The study of deeper aquifers needs geophysical surveys, bore-well logging *etc*.

Remote Sensing

The population growth has been creating more and more stress on agriculture sector for increasing the food grain production, which consequently increased deforestation and demand for more water. The available surface water resources are inadequate to meet the entire water requirement for various purposes. So the demand for underground water has increased over the years. Generally, groundwater is less prone to pollution in comparison to surface water. Hence, groundwater serves as an important source of water for various purposes in rural and urban areas. In recent years intensive use of satellite remote sensing has made it easier to define the spatial distribution of different groundwater prospect classes on the basis of geomorphology and other associated features (Kulkarni 1994; Adyalkar 1996; Goswami *et al,* 1996; Srinivasa Rao *et al,* 1997; Bhan 1998; Bhagwan 1998; and Patil *et al,* 1999).

In many earlier studies (Karanath and Seshu Babu, 1978; Raju *et al,* 1985; Satyanarayana 1991; and Srinivasa Rao *et al,* 1997) remote sensing techniques have been applied for groundwater prospecting.

Remote sensing with its advantages of spatial, spectral and temporal availability of data covering large and inaccessible areas within short time has become a very handy tool in assessing, monitoring and conserving groundwater resources. Satellite data provides quick and useful baseline information on the parameters controlling the occurrence and movement of groundwater like geology, lithology/structural, geomorphology, soils, landuse/cover, lineaments etc. However all the controlling parameters have rarely been studied together because of non-availability of data, integrating tools and modelling techniques. Hence a systematic study of these actors leads to better delineation of prospective zones in an area, which is then followed up on the ground through detailed hydrogeological and geophysical investigations. Visual interpretation has been the main tool for evaluation of groundwater prospective zones for over three decades. It has also been found that remote sensing besides helping in targeting potential zones for groundwater exploration provides inputs towards estimation of the total groundwater resources in an area, the selection of appropriate sites for artificial recharge and the depth of the weathering area.

By combining the remote sensing information with adequate field data, particularly well inventory and yield data, it is possible to arrive at prognostic models to predict the ranges of depth, the yield, the success rate and the types of wells suited to various terrain under different hydrogeological domains. Based on the status of groundwater development and groundwater irrigated areas (though remote sensing), artificial recharge structures such as percolation tanks, check dams and subsurface dykes can be recommended upstream of groundwater irrigated areas to recharge the wells in the downstream areas so as to augment groundwater resources.

Apart from visual interpretation, many researchers for deriving geological, structural and geomorphological details use digital techniques. The various thematic layers generated using remote sensing data like lithology/structural, geomorphology, landuse/cover, lineaments etc, can be integrated with slope, drainage density and other collateral data in a Geographic Information system (GIS) framework and analysed using a model developed with logical conditions to derive at groundwater zones as well as artificial recharge sites. The GIS for groundwater targeting, management and conservation of groundwater resources that ensures optimum and judicious use of groundwater and in identification of artificial recharge sites is described in next section.

Aerial Photointerpretation

Aerial photography, photogrammetry and aerial photointerpretation are the recently developed techniques having the close association with the remote sensing technique. An *aerial photograph* is the photographic image of the earth's surface usually taken from the aircraft. The *aerial photointerpretation* is an act of examining the aerial photographic images for the purpose of identifying objects and judging their significance. *Photogrammetry* is the art of making reliable measurements from the aerial photographs. Some important aspects required for aerial photointerpretation are given below:

1. Nature of Photogeology

The use of aerial photographs to obtain both qualitative and quantitative geologic information is referred to as photogeology.

2. Photogeology Uses

Geologists commonly use aerial photographs for the following purposes:

i. Compiling topographic and geologic maps.
ii. Exploring for mineral and fuel deposits, and finding groundwater reservoirs.
iii. Identifying hazardous sites such as earthquake and landslide zones.
iv. Mapping landscape changes due to natural hazards (hurricanes, floods, *etc*.)
v. Geotechnical investigations, selecting construction sites for buildings, roads, and dams.
vi. Soil surveys and forest inventories, *etc*.

3. Recognition Elements

Shape, size, pattern, shadow, tone or color, texture, association, and site are all used to recognize photogeology features. The quantity and quality of geologic information that can be interpreted from aerial photographs is dependent on the following factors:

i. The type of terrain, vegetation and soil cover, and the stage of erosion.
ii. The type and scale of the photography.
iii. Availability of stereopairs and the amount of vertical exaggeration present.
iv. Training and experience of the photo interpreter.

4. Sun Angles

Most photogeology studies are accomplished using photos obtained with high sun angles to reduce affects of shadow. There are special times; however, when low sun angles and shadow enhancement is of value, particularly in areas of low relief. Aerial photography should normally be confined to the period when sun angle is more than 30^0 but less than half the angular convergence of the lens. The optimum sun angle is 45^0.

5. Convergence of Information Principle

This principle is often used whereby an interpreter starts with small-scale photographs for a larger (synoptic) view of the land and gradually focuses upon larger and larger scale photographs to target features of specific interest. This is augmented with fieldwork to verify the discoveries and to provide for detailed analysis of the site.

6. Scale of Photograph

The scale of photography depends upon the objective of the work, which governs the speed of aircraft and its flight height. Scale of photograph is the ratio of focal length (f) of camera and the flight height (H) *i.e.* scale= f/H.

Qualitative and Quantitative Information from Aerial photos

Qualitative Information

Four major categories of qualitative geologic information are available on aerial photos:

1. Rock Type (Lithology)

Lithology is explained in terms of rock origin as igneous, sedimentary, or metamorphic rocks. Landforms composed of these rock classes often show up with striking clarity on aerial photographs.

The outcrops of various kinds of rock that have been exposed through weathering and erosion can tell a great deal to photo-geologists about the composition of the rock units and their strike and dip relationships. As climate becomes more humid, many of these lithologic relationships are hidden under soil and vegetation covers that make their identification more difficult. The photo interpreter uses a combination of topographic expression, drainage pattern and texture, residual soil tone, structural imprints, and the zoning patterns of natural vegetation for aerial photo-interpretation.

(a) **Igneous** - Igneous surface materials that may be directly recognized by inference include granite, basalt, lava flows, volcanic glass, and pumice. Light photographic tones often indicate areas of granite, rhyolite or tuff, while dark tones express the presence of basalt, gabbro or andesite.

(b) **Sedimentary** - It may be possible for the interpreter to recognize gypsum sand, quartz sand, sandstone, dolomite, and limestone deposits. Sandstone commonly appears as light tones on panchromatic photos because of the characteristic development of a thin mantle of sand. Forming more closely spaced dendritic stream patterns then sandstone, shales photographic tones are mottled due to variations in moisture but are often darker than those of sandstone. Limestone are generally light toned on photographs but can be easily identified on the aerial photographs due to presence of sink holes.

(c) **Metamorphic** - Gneiss is the principal metamorphic rock that is commonly observed and marble may be detected in open quarries. Of limited extent worldwide, metamorphic rocks may display some of the elements of sedimentary rocks (banding and alternating ridge and valley, or stair-step valley walls) if the metamorphosed sediments are of different resistance to weathering and erosion. If formed from the igneous rocks, igneous rock features (uniformity of color and dendritic drainage patterns) may predominate.

2. Rock Structure

Structural features denote geologic structures that are formed by internal deformation forces in the earth's crust. Flat lying, tipped, folded, and faulted structures of the earth's crust produces distinctive patterns in the rocks that are discernable on aerial photographs.

3. Drainage Patterns and Texture

Drainage patterns represent the arrangement and repetition of stream channels that have been formed in response to natural forces acting upon the earth's land surface. The patterns reflect to a varying degree the lithology and structure of a region, and the nature of the erosional and depositional forces active in moving sediment. Drainage patterns also reflect the erosional stage that the landscape is in. Climate often controls the amount of moisture, soil formation, the depth of weathering, the amount of vegetative cover, and the rate of erosion and deposition in an area. It in turn affects the stage of erosion and thus appearance of a landscape.

Drainage patterns can be further classified by variations in channel density per unit area. Three main types of drainage texture are observable on aerial photographs:

1. Fine-Textured Drainage - Created by the development of many closely spaced tributaries on loose or poorly cemented clay and silt sediments of low permeability.
2. Medium-Textured Drainage - Moderate density of stream channels that develops on soil and bedrock have a moderate permeability.
3. Coarse-Textured Drainage - A low drainage density that develops on hard, resistant rock formations or highly permeable materials.

4. Landforms

Landforms are the distinctive geometric configurations of the earth's surface. Landforms may be defined as constructional (volcano, coral reef), erosional (stream valley, mesa), or depositional (floodplain, glacial moraines). Most surficial deposits (unconsolidated sediments) can readily be distinguished from consolidated rocks on aerial photographs. Those materials that were transported by water (fluvial), ice (glacial), wind (aeolian), and gravity (mass wasting) often have distinctive forms.

Quantitative Information from Aerial photos

It is possible to make several types of direct geologic measurements on aerial photos:

1. Size, shape, pattern and density measurements
2. Strike and Dip Determinations - Patterns of fractures, faults, and intrusive dikes are usually discernible on aerial photographs. Their strike direction can be measured by using compass bearings or azimuths. Dip is the angle of inclination, or the amount the surface is tilted from a horizontal plane. The dip is always measured at right angles to the strike. When bedding surfaces coincide with topographic surfaces, such as in flat-lying sedimentary rock areas, the dip angle can be determined using parallax by measurement of the height difference between two points, one directly downslope from the other. After the horizontal distance between the same two points is determined, the dip angle can be computed by this trigonometric relationship:

   $$\text{Tangent of dip angle} = \frac{\textit{Vertical distance}}{\textit{Horizontal distance}}$$

   If relief in an area is low, the horizontal distance may be scaled directly from a single photo without significant error in computation of the dip. When relief is higher, a correction for the relief displacement of the upper point with respect to the lower point should be made.
3. Identification of Fault Zones on Aerial photos - A number of clues are available for locating possible faults on aerial photos:

(a) Lineaments - Linear features (lines) often appear on aerial photos in regions of little soil and vegetation that can be attributed to faulting of the bedrock. Lineaments often occur in fields of linear cracks often oriented in the same direction.

(b) Deflected or unusually straight stream channels.

(c) Straight contacts between erosional and depositional features.

(d) Straight valleys in hard rock areas.

(e) Linear alignments of natural vegetation and lakes, including sag ponds.

(f) Topographic scarps.

(g) Distinct hue or tonal changes on opposite sides of a lineament.

Even though remote sensing and aerial photograph have wide advance over conventional techniques of hydrogeomorphology, but may have limitations in the form of practical miss-conclusions. Hence after generating the thematic maps by using these techniques the field visits are essential to come a particular conclusion.

Geographical Informational System (GIS)

Geographical Informational system (GIS) is a computer-assisted system designed to capture, store, edit, display and plot geographically referenced data. It has the capability to make quick and unbiased decision, which includes distance, directions, adjacency, relative locations and other spatial concepts. This new wave of information technology is relevant to organizations like Government, Corporation, Municipality, Industries, *etc.* which are involved with resources planning and management of facilities, designing and management of networks, monitoring, assessment, development and planning for the future, etc (Arya *et al,* 2002, Bhushan 2002, Gupta *et al,* 2002 Jain and Reddy 2002, Jenson and Dominique 1988, Kumar *et al,* 1996, Maji *et al,* 2002, Mote and Shekdar 2002, Nayak 2002, Ravishankar and Venkataratnam 2002, Sara Naaz and Venkateshwara Rao 1998, Sharma *et al,* 2002 and Vazir Mahmood and Durga Rao 2002).

Here are few examples of application of GIS techniques:

- Natural resources assessment, mapping, monitoring and management - forestry, water resources, agriculture, urban and rural development, disasters, *etc.*
- Designing/managing networks and distribution such as electric, telephone, *etc.*
- Managing network of pipes in oil and gas industries, refineries, chemical plants, iron and steel plants and others.
- Planning / managing a city, corporation or municipality water and sewer distribution systems.

There are three types of GIS based on whether the information system has to provide geometric analysis, attribute analysis or spatial analysis Burrough, (1986).

Correspondingly, the systems associated with such principal areas are:

- Computer Aided Mapping (CAM)
- Automated Mapping and Facilities Management (AM/FM)
- Geographic (topology) Information System (GIS)

CAM: The CAM employs Computer Aided Design and Drafting Technology (CADD) primarily intended for map production. This technology replaces the traditional manual cartographic process that has been used and is still being used in certain places for the production of maps. CAM is not suitable for spatial data analysis and answering queries because the spatial relationships are not defined in the data structure, an essential component in GIS application.

AM/FM: AM/FM system as the name suggests, refers to automated mapping and facilities management developed based on CADD technology suited to the needs of facilities management, more specifically, network distribution, be it electrical, telephone, pipeline and transportation. In this system, the emphasis is laid on storage, analysis, and reporting of both graphic and non-graphic data that have geographic reference. The most important feature of the system is that the non-spatial attribute data network is linked to graphic data.

With the support of sound database, following are some of the benefits that can be derived from AM/FM system:

- Efficient use of the resources in network.
- Network load balancing.
- Design/redesigning of the network and appropriate selection of components.
- Network capacity assessment.
- Maximization of utilization.
- Automatic generation of work orders for routine maintenance of network/installation target.

GIS (Topology): The GIS built around topology is ideally suited for analysis of geographic data, which has the spatial relationships among all the data elements. The topology in this system describes not only the location and geometry of the map features, but also provides information on linear map feature, connections, boundary and its contiguity.

The database in GIS comprises both spatial and non-spatial data with attribute data attached to them. The software used in the system permits routine analysis of both spatial and attribute data simultaneously with the result user is able to search the attribute data and relate that to spatial data and vice versa.

Today, much commercially available software is capable of being used as either GIS or AM/FM system. Usually such software is modularly organised such that each module provides a distinct application capability.

A core module is usually provided by the manufacturer, which can be utilized with the corresponding application module, specifically formatted, for a chosen application. A core module generally deals with the data base organization, linking of spatial or non-spatial data, security features and records management. The application module contains specific analysis routines that are needed for the user to efficiently manage the system *viz.* land information system, water scheduling system for irrigation, forest information system, disaster management system, *etc.*

Water, one of the most essential material in day-to-day life is becoming scarce due to various reasons including reduction in infiltration rates, higher rates of runoff, uneconomical use, overexploitation of the surface resources etc; as a result of change in land use patterns, degradation of forest cover and public apathy towards its importance. An understanding of the complex inter sectoral dynamics would be crucial for developing a holistic approach to the utilization of water resources (Maidment, 2002). For managing the data at basin level and analysing the data correlation between the various sectors in the basin, GIS has been found to be an effective tool. People have varying goals and values relative to use of local land and water resources, which need to be properly managed using the GIS techniques.

Geographic Analysis/Modelling:

The goal of geographic analysis and modelling in GIS environment is to find answers to physical issues related to the application: in this study, to identify hot spots in the study area and depict scenarios for hypothetical what-if questions. Analysis and modelling of geographic data may be on either of the two data domains integrated by GIS systems, namely spatial or geometric data and attribute data domains (Sridhar *et al,* 1994).

The effectiveness of a GIS in solving application problems is highly dependent on the techniques used and the kind of operations supported by the GIS package. In addition to conventional vector overlay operations; a number of other techniques can be used for analysing and modelling geographic data. For any project a number of GIS techniques identified are to be used. The following techniques can be used for this purpose:

- GIS theme operations
- Formula application
- Qualitative analysis
- Change detection analysis
- Socioeconomic data models
- Influence zone model
- Multi-parameter analysis
- Presentation charts

Analysis and modelling operations involve multiple steps creating data for the next step. Due to the size of data, operations can be time consuming.

The above techniques of geographic analysis and modelling employed are explained below briefly.

(i) GIS theme Operations

A GIS supports a number of operations on thematic layers for combining, extracting and transforming map data. In vector GIS packages these operations work on both spatial and attribute data domains (Sridhar *et al,* 1994). The operations are well-defined geometric and topological operations, which are also given semantics in the attribute domain. Due to the evolutionary nature of GIS systems, the attribute domain semantics is somewhat ad-hoc. For some of the theme based operations, which combine thematic layers, the input layers used for the operations must be geo-referenced. GIS theme operations are sometimes referred to as spatial modelling or overlays (Burrough 1986, Gupta *et al,* 2002 and Nayak 2002).

(ii) Formula Application

An attribute column value in the data tables of the geographic database may be computed by a mathematical formula using some of the other data columns in the table. For example, the watershed wise groundwater recharge can be computed as a formula based on area value and average rainfall of a land parcel (Vazir Mahmood and Durga Rao 2002).

(iii) Qualitative Analysis

A typical method used in GIS modelling is to compute numerical values for each spatial feature in a theme and classify the numerical values on an interval basis. For example, the literacy levels of blocks may be ranked on a five-point scale of values very good, good, average, poor and very poor based on numerical values of literacy for each block. This method converts quantitative data to qualitative data depends on the domain knowledge of the expert choosing the intervals for quality classification. A number of statistical techniques can be used for the classification (Burrough 1986).

(iv) Change Detection Analysis

An important use of a GIS is to identify temporal changes in spatial data. The temporal factor may be based on past, current and future scenario data in the geographic database. For example change in forest cover over decades can be detected subject to availability of forests cover data for decadal periods. Change detection analysis is done using spatial modelling operations such as theme combination, extraction and polygon aggregation.

(v) Socioeconomic Data

Models Socioeconomic data modelling primarily works on attribute data and the result of the modelling is associated with a suitable spatial feature. In modelling of socioeconomic data it is necessary to calculate a functional hierarchy of settlements (Sharma, 1981) in the study area so that they may be compared to assess the relative socioeconomic development of the settlements chosen for the study. A composite functionality index is determined for each unit of settlement (SAC, 1992 and Sridhar *et al,* 1996) as a comparative index. The development index may itself be defined on multiple parameters such as availability of power, transport facilities, communication facilities, *etc.*

(vi) Influence Zone Models

Influence zone models convert spatial data in point, line or polygonal form to polygonal areas representing influence zones. Examples of such influence zone modelling techniques are constructed of buffer corridors, generation of voronoi diagrams (Thiessen polygon) (Auren 1991) from a triangulation model and isolines generation from a triangulation model. A buffer corridor around a mining area can approximate the likely spread of mining or building activities; a voronoi diagram constructed on rainfall data measured at point locations can depict the average rainfall collected around the measuring station.

(vii) Multi-Parameter Analysis

A recurrent theme in GIS modelling is multi-parameter analysis. A number of GIS applications need to consider multiple parameters to identify suitable areas that meet the required goal. For example, to identify agriculturally suitable areas parameters such as land capability, soil slope, geomorphology, *etc.* need to be taken in to account. Subjectivity in the definition of relative importance of the multiple parameters for the goal is inevitable.

Multi-parameter analysis may be done by one-domain experts who examine the values for the parameters considered and arrive at a relative scale of suitability. The importance matrix method of (Saaty, 1980) for multi-prameter analysis is a technique to isolate the subjectivity in the analysis and has been used in (SAC 1992 and Sridhar et al, 1996).

(viii) Presentation Charts

Presentation charts such as pies, bars, graphs *etc.*, provide an easily understandable graphical method of depicting tabular data. In a GIS environment such charts can be spatially associated, or geo-referenced, with features in a map. For example the crop production of all districts can be presented as pie charts superimposed on each districts of the map.

Methodology of GIS Applications

- In order to demarcate the groundwater potential zones of study area different thematic maps on 1:50,000 scales are to be prepared from remote sensing data, topographic maps and resistivity data.
- Drainage map is to be prepared from SOI toposheet and satellite data.
- The thematic maps on hydrogeomorphology and lineaments are to be prepared using IRS satellite imageries data by visual interpretation on 1:50,000 scales
- Contour map and spot elevation map are to be prepared from SOI toposheet.
- All primary input maps (hydrogeomorphology, lineament, contour and spot elevation, drainage and geo-electrical sounding location) are to be digitized in Arc/Info, GIS software package and slope map is to be made from digital elevation data.
- Interpretation of geo-electrical soundings data.
- Correlation of geo-electrical parameters of drilled sites with lithology.
- Based on above co-relation lithology is to be inferred at other sounding locations for identifying horizontal and vertical variation in subsurface lithology and estimating depth to the hard rock.
- Using inferred lithology and thickness from geo-electrical parameters at respective locations; aquifer layer thickness and overburden thickness maps are to be prepared through GIS.
- The different polygons in the thematic layers are to be labeled separately and then they will be registered. In the final thematic layer initially each one of the polygons are to be qualitatively visualized into one of the categories like (i) very good (ii) good to very good (iii) good (iv) moderate and (v) poor in terms of their importance with respect to groundwater occurrence and suitable weights can be assigned.

- Finally thematic layers are to be converted in to grid with related item weight and then to be integrated and analysed, using weighted aggregation method. The grids in the integrated layer are to be grouped into different ground water potential zones by a suitable logical reasoning and conditioning. The final ground water potential zone map thus generated must be verified with the yield data to ascertain the validity of the model developed.

In addition to ground water data processing, maps generated from various GIS platforms can be integrated with "GWIS" for spatial representation and visualization. Various location and thematic content maps become an integral part of the ground water database. The exchange of information with other software package is enabled via import and export options.

The methodology of GIS application is given in Fig. 10.1.

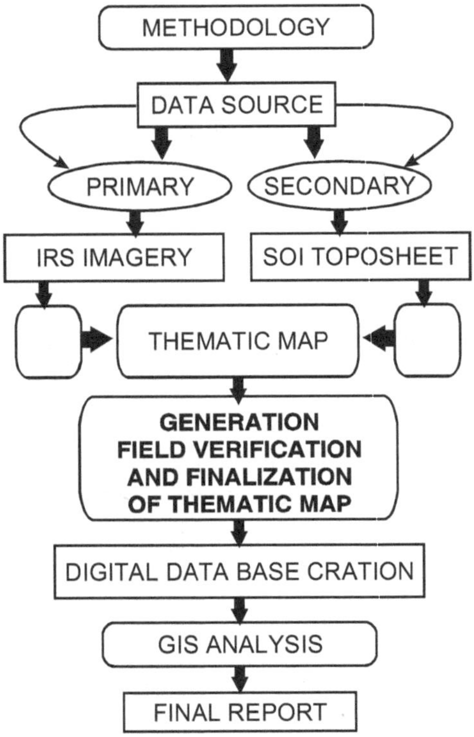

Fig. 10.1 Schematic diagram of GIS Application Methodology

Global Positioning System (GPS)

A system for providing precise location, which is based on data transmitted from a constellation of 24 satellites, is known as Global Positioning System (GPS).

So far geological investigation of the site using traditional methodology have not yielded better results. Many projects could not stand the test of time either because proper geological considerations were not taken care off or due to faulty geological investigation. Now days, geological considerations are given due recognition in the construction of engineering projects. The traditional methodology alone is no longer followed, as the chances of error are just too much in them. The analysis of geological parameters has not only become fast but also highly accurate in modern time involving use of recent techniques.

Modern investigation techniques, such as GPS, GIS and Remote Sensing techniques can provide fast and desired accuracy to surface and subsurface parameters. Global Positioning Systems are capable of recording positions to a high level of accuracy. GPS positioning system is probably the most effective general-purpose tool created to date for the determination of movement, spatial specifications, track history and time (Kandasamy and Venugopal 2002, Kulkarni *et al,* 2002, Prasanna 2002 and Wegner 1999). GPS system as a whole can be divided into three segments:

- **Space segment:** Constellation of satellites
- **User segment:** GPS receiver, hand held or vehicle installed
- **Control segments:** Unmanned ground station to monitor the satellites

The concept behind GPS is the distance measurements between the receivers and satellites by the method of space resections. Differential GPS is required to obtain the accuracy required by many GIS data capture applications in terms of sub-centimeter scale. DGPS requires the use of a base station at a known location to remove systematic error from the GPS signals. Recent GPS / GIS data collection devises have the capability to collect a wide range of GIS attribute data in addition to position of interest such as points, lines and area features.

Taking into consideration, the inhospitable terrain, where most of the engineering constructions are carried out by using GPS supplemented by GIS and remote sensing techniques by a hydrogeologist, is highly recommended. GPS and remote sensing techniques are very useful tools for data capture. GPS in fact is highly useful to cross check the data collected by a satellite imagery analyst and the surveyor.

GIS technique is basically a data analysis and display tool. GIS is a computer based system, which is an exceedingly useful tool to analyze the hydrogeomorphological features present on the surface of the earth and various spatial changes taking place on it with respect to time. In major engineering projects, such as dams, bridges, tunnels, high relief roads, data collection is a very difficult job, due to the inhospitable environment in most of the construction sites. Hence, taking into account such hazards or any unforeseen circumstance, GPS and remote sensing technologies will provide not only fast and accurate but also reliable hydrogeomorphological and geological data, which can be very successfully analysed by GIS, to give complete picture about the stability and durability of the structure.

The advantages of GPS/remote sensing techniques/GIS can be summarized as below:

Multifunctional: apart from the location data collected by GPS, the other attributes, which can be recorded, are:

- Traverse distance
- Bearing of a station
- Track history
- High precision clock
- In built memory for data record
- Net displacement
- Datum option
- Operational simplicity
- Portability
- Compatibility with modern electronic gadgets

GPS used in association with GIS may help on-site data capture, its analysis and its output. GPS with hand-held range finder, especially in case of inaccessible. Verification of location data collected from satellite imagery by GPS.

Digital Elevation Model (DEM)

Digital elevation models (DEMs) are generated using satellite imagery and used in the study of hydrogeomorphological products such as relief, elevation contours, basin boundaries, stream networks etc. Automated extraction techniques for watershed hydrological studies often begin with the generation and analysis of basin hydrogeomorphological products, including basin boundaries, stream networks and slope aspect of maps. These hydrogeomorphological products are traditionally derived from watershed topographic maps, and there currently exist a suite of algorithms that automatically extract basin hydrogeomorphology from digital elevation models (O'Callaghan and Mark 1984; Band 1986; Jenson and Dominique 1988; Tarboton *et al.,* 1991; Helmlinger 1993, Martz and Garbrecht 1993; Moussa, 1997, Tarboton, 1997 and Endreny *et al,* 2000). Slope and relief aspect can be calculated for each of the post-processed DEMs to determine the cell drainage direction, which was represented as a one-dimensional ¯fow vector towards the lowest of the eight neighboring cells, as described by O'Callaghan and Mark (1984). After slope and relief aspect can be computed, the directions and accumulation routines were used to identify drainage networks and a watershed boundary can also be computed by including all cells from the DEM that drained to (the pixel at) USGS gauge 07327550 (Endreny *et al,* 2000).

The broad methodology used to create digital data for DEM thematic layers of the macro database is the following :

1. Digitizing of individual map sheet tiles
2. Construction of polygon topology wherever required
3. Editing of digitized themes
4. Association of names/attribute data with spatial features
5. Geo-referencing, mosaicking & edge matching of individual tiles
6. Clipping with DEM boundary
7. Symbolization

Steps (1) to (4) complete the creation of digital data for the individual tiles. Step (5) merges the individual tiles to create a composite thematic layer for entire area.. Step (6) eliminates spatial features outside the DEM area in the composite map. The final step (7) symbolizes the map data by suitable coloring and symbol association.

Grid-based digital elevation models (DEM) are an array of elevation values that provide an estimate of terrain surface useful in computer-based hydrogeomorphological analysis, such as the automated extraction of topographic relief, elevation contours, basin stream networks and basin boundaries (O'Callaghan and Mark 1984; Band 1986; Moore *et al.,* 1991). Stereo pairs of satellite images, when processed with photogrammetric techniques (Helpke 1995), provide DEMs that can rival the resolution, geographical reach and cost of traditional high altitude photogrammetric DEMs. A global set of DEM products currently exists for hydrogeomorphological analysis, but their resolution and coverage vary considerably (Table 10.1).

Table 10.1. DEM data set availability, resolution and coverage by country*

Country	**Resolution (m)**	**Coverage (%)**
Australia	100	100
Belgium	30	100
Canada	100	29
Finland	25	100
France	50	100
Greece	250	100
Israel	25	100
Italy	20	50
Japan	50	40
Malaysia	15	10
Netherlands	10	100
New Zealand	25	100
Portugal	10	60
South Africa	400	0.5
United Kingdom	50	100
United States of America	30	58
Vietnam	1000	100

* Jet Propulsion Laboratory (1997)

In basin-wide hydrogeomorphological assessments, which included examination of Horton ratios and the general shape of basin boundaries, the SPOT-derived DEM and its associated geomorphological products had few differences with the higher accuracy USGS, USDA and EPA products. As such, meso- and macro-scale research in basins beret of high accuracy maps (Table 10.1) would most definitely gain a great deal from use of satellite-derived imagery with error structures similar to SPOT products. SPOT-derived DEMs used to perform a SPOT panchromatic image ortho-rectification, a common procedure for SPOT data, would enjoy this same accuracy at meso- and macro-scales. At the micro-scale, however, the presence of 100 m average differences between the SPOT products and the associated higher accuracy product forces a great spatial uncertainty on many geomorphological predictions (Endreny et al, 2000). Whether the product was a contour line image, a prediction of terrain slope and aspect assumption for a stream network, the SPOT satellite data did not provide a close match to the proven and higher accuracy USGS level 2 maps. Exercises in scaling revealed that these spatial discrepancies were quickly constrained at distances of 500 m to 5 km, a small distance compared with scales used in many of the 2-, 4-, 6-, for 8-digit HUC hydrogeomorphological studies (USDA, 1981).

Groundwater Information System (GWIS)

For Groundwater Information System (GWIS) the two independent packages namely GRAM++ (DST-CSIR, 1999) and GWW (Braticevic and Karanjac, 1997) are useful. GRAM++ is window based GIS package, which can be used for analysis of spatial and non-spatial details such as village boundaries, communication, land-use, socio-economic data etc. and related quires (Kumar *et al,* 2002). The GWW can be used for analysis of geohydrological parameters such as master data, chemistry, pumping test, hydrographs, mapping, well log, cross section, fence diagrams, step draw down test, grain size curve, and abstraction *etc.*

The following themes have been taken into consideration for GWIS in GRAM++: Village boundary and demographic characteristics, drainage, transportation network, land-use, hydrogeomorphology, surface water, topography, groundwater development feasibility *etc.*

The following parameters have been taken into consideration for establishment of GWIS in GWW:

1. Location parameters for wells, sampling points, springs etc. (Well identification code, SOI toposheet code, owner, use of well, village, block and district name, X and Y coordinates, Ground surface elevation, regional belonging (basin/ watershed), geology, aquifer types, types of well)
2. Water Level Data: new observation wells, observation wells of CGWB and state groundwater boards.
3. Lithology, stratigraphy
4. Hydro-geological and hydrogeophysical Parameters aquifer characteristics, depth and thickness of aquifer, effective porosity, aquifer resistivity. Transmissivity, hydraulic conductivity and leakage coefficient of aquifer has not been included due to non-availability of sufficient data.

The following are the data retrievals and presentations using the database and information system created with the GWW software

- Location maps showing points (observation well, vertical electrical sounding sites, surface elevation, basement depth observation location), lines and areas (watershed boundary)
- Various contours maps (surface elevation, depth of basement, water level, aquifer resistivity)
- Lithological cross-sections (in present study it is electrical log)
- Fence diagrams showings the lithology, stratigraphy and water levels or heads

The Geographic Information System (GIS) and Groundwater for Window (GWW) has helped in creating interactive groundwater information system. The scattered non-standard data on different parameters available with different users agencies have been well arranged in retrievable format in GWW. The GWIS of Upper Barakar basin is providing well-arranged and structured database (Kumar *et al,* 2001). The created information system in GIS and GWW may be used to the standard groundwater modelling package. However due to lack of interface between GIS package and GWW, the data related to other natural resources cannot be actively linked together. There is further scope for development of interface between GIS (in present case GRAM++).

The approach adopted in the creation of GWIS for Upper Barakar Basin, Hazaribagh (Kumar et al, 2001) may be used in the other parts of Country. GWIS of Upper Barakar has provided opportunity to understand the different themes of groundwater in integrated manner.

Digital Surface Terrain Modelling (DSTM)

The lineaments incorporated in the hydrogeomorphological map are supposed to be the probable fractures. Hydrogeomorphological mapping and lineament analysis using remotely sensed data has emerged as a new tool. But at the same time, numerous case studies came out (Kumar *et al,* 1996, Kumar and Prasad 1993 and 2002) revealed that this new emerging tool has serious limitations and technology has been over used. But one cannot deny that aquifer is not a planer thing rather it has 3-D characteristics (aquifer geometry). Similarly remotely sensed lineament is a planer thing whereas fracture has 3D aspects. Therefore if some one is thinking for sustainable development, utilization and management of groundwater in watershed prospective, they must take into account of climatic, physiographic and aquifer geometry characteristics in hard rock area.

In this study the results of surface topography through Digital Surface Terrain Modelling (DSTM) and aquifer geometry analysis will be highlighted. Further scope and utility of DSTM will be considered as a tool for sustainable utilization, development and management of groundwater in the area.

Following methodology can be adopted for DSTM study.

- Analysis of geo-hydrological and rainfall data,
- Digitization of topographic (elevation) data of the watershed for generation of DSTM
- Electrical resistivity (vertical electrical sounding) on different landforms as well as sufficient number of vertical electrical sounding (VES) necessary for depicting the regional and site-specific basement depth fractures of the watershed.

- Generation of hydrogeophysical parameters of individual VES sites.
- Generation of DSTM keeping constant terrain module factors for the watershed.
- Identification of fracture system, sub-surface basins for aquifer storage and retrieval.

Multi-Criteria Evaluation (MCE)

Multi-criteria evaluation (MCE) techniques are numerical algorithms that define the suitability of a particular solution on the basis of the input criteria and weights together with some mathematical or logical means of determining trade-offs when conflict arise (Heywood *et. al.,* 2003). In this technique, 'weight' is assigned to the data layers to reflect their relative importance. Groundwater exploration in a geologically complex terrain requires consideration of a number of factors – both natural and anthropogenic. It is important to understand the control of these factors on the groundwater regime of any area for optimal exploitation and aquifer management particularly in and around urban growth centers. As such, to arrive at a clear picture of the situation, the controlling factors have to be treated and integrated giving weight that is specific to a particular area. In this way the MCE technique used in GIS environment is a framework for decision-making problems pertaining to ground water exploration.

The Database for MCE is the Survey of India toposheet maps in the scale of 1:50,000 used to prepare the reference map for the study area with supplementary information derived from satellite imageries (*e.g.* IRS IA, Geocoded FCC, IRS ID, LISS III, PAN sharpened, *etc*). The database for geology and geomorphology can be prepared from satellite images in combination with fieldwork and input from available literature on the subject. With the help of this technique different geomorphic units *viz.* the denudo-structural hills (residual hills), the alluvial plains and the marshy lands including the static water bodies (Water bodies with paleo-channels), *etc.* can be identified supported by satellite images and toposheet maps.

Thermal Infra Red (TIR) Mapping and Microwave Data

Sub-surface investigations for geological data has got a boost up in the light of use of electromagnetic radiation in the micro wavelength region (1 mm - 0.8 m) and thermal infra red (1 micro m - 1 mm.) region. Microwave remote sensing has a unique advantage of penetrations into the sub-surface and it is able to provide information on soil moisture, snow cover, surface condition below vegetative cover. Here we have two major types of sensor:

- Active microwave remote sensor
- Passive microwave remote sensor

Active microwave remote sensor illuminates the ground with microwave radiation and then receives the back-scattered energy from the object. Some of the active microwave remote sensors are:

- Side looking airborne radar (SLAR)
- Synthetic aperture radar (SAR)
- Wind scatterometer
- Altimeter
- Polarimeter
- Rain mapping radar

Passive sensors are called microwave radiometers, which receive and detect the radiation emitted from various objects on the earth.

In the TIR region the thermal and reflected properties vary and their intensity changes with the change in lattice structure and mineral compositions. Moreover, lithological mapping of an area can be carried out by the measurement of Apparent Thermal Inertia (ATI) of a rock, which is different for different litho types. The ATI signature of a rock is especially affective in the night time as the effect of solar radiation is minimum during nights. For example, limestone and dolomite delineation is difficult in visible range of electromagnetic radiation, but dolomite having higher thermal inertia compared to limestone can be distinguished by night time thermal image. Different types of geological structures such as anticline and syncline also show warmer and cooler signatures. Similarly, structural discontinuity such as fault and fracture also show cool linear anomalies in both day and night time, hence making their delineation possible.

Side-Looking Airborne Radar (SLAR)

SLAR images are acquired by sending a beam of radar energy to the ground at an angle perpendicular to the aircraft's flight path. With

SLAR the terrain is illuminated at an oblique angle to enhance subtle geologic structures such as folds and faults. Light and dark areas on the image are caused by high and low radar reflectivity, respectively. The SLAR images can be taken by the flight at a flight altitude greater than 8,000 meters above mean sea level, on 30- 45 km-wide strips flown along 20 km flight lines. The STAR-1 Synthetic Aperture Radar System should be used in wide swath mode with the particular look direction east. Sections from the image strips must be mosaiced and photogrammetrically controlled to 1:250,000-scale *i.e.* Universal Transverse Mercator (UTM) topographic maps of the area.

Lineament interpretation in this study is based on the criteria used by Lattman and Parizek (1964), and as described by O'Leary *et al.* (1976). The lineaments can be mapped on overlays from individual SLAR swaths, accompanied by a 1:240,000 SLAR mosaic. This overlay is then digitized and can be imported into ARC/INFO 7.2.1™ software for editing. Arc View 3.0b™ software will be then used to create the lineament map.

SLAR images are also useful to create the geologic map. The geologic unit boundaries apparent on the SLAR swaths can be mapped and then compare it to previous geologic and hydrological maps of the area at various scales. A lithological class can be assigned to each mapped unit based on a review of the previously published maps. A unit boundary mapped from the SLAR often corresponded well to a unit found on a previously published map, but not always the same map. Geologic mapping using the SLAR imagery results in a map that contains a group of units not found on any one previously published geologic maps.

Lineament frequency and occurrence derived from the SLAR swaths can be explained by the surrounding geology, and can help clarify some of the geologic questions. In areas with thick surficial cover, for instance, there are few lineaments, but those lineaments that are apparent from fluvial systems or karst topography may help to explain some of the tectonic history covered by the sediments.

The geologic map based on the SLAR correlates well with geologic maps produced by onsite field mapping and compiled at larger scales. Some map units that can be clearly visible on the SLAR are not found on some geologic maps, but included on others. Other mappable features (such as lineaments) on the SLAR are less readily apparent from the field. Also, in many instances the use of SLAR imagery could be more efficient than mosaicking and interpreting a large quantity of aerial photographs.

The utility of SLAR for geologic mapping in any area lies in the ease with which both structure and lithology are discernable. Unlike visible and near-infrared wavelengths, radar energy (at the correct wavelength) can penetrate most clouds, making it an especially useful tool in the regions where a persistent cloud cover generally obscures parts of the study area. SLAR can also be used to supplement other forms of data, such as aerial photographs and satellite imagery, which are limited by cloud cover. This type of study of SLAR of any particular region would be useful on other tropical regions with the same difficult conditions (vegetation, rugged terrain, cloud cover and poor roads).

Chapter 11

Hydrogeomorphological Mapping

≈≈≈

⇒ This chapter deals with the guidelines for hydrogeomorphological mapping, generation of thematic maps geomorphological map, landuse map, soil map, groundwater potential, digital map and hydrogeomorphological action plan map.

Common groundwater investigations include necessary considerations of the system's influx sources, flow regime, discharge points and geomorphological characteristics of the area. Often, hydrogeomorphological mapping of an area is required as to determine these characteristics. The primary data can be acquired from the survey of India toposheet map and the satellite images. The secondary data set includes the satellite imageries, the preparation of digital elevation model (DEM) and digitized precipitation maps.

The first step in the procedure of hydrogeomorphological mapping is to calculate the watershed/basin area, which serves as a mask to define the area of interest. The watershed/basin calculation relies on the digital elevation data and assumes that topographic aspect (dip direction of the cell surface) completely controls the extent to which surface flow emanating from each point in the study area can reach the low point in the valley. The calculation of the potential annual recharge of the aquifer combines techniques described above with data for recharge areas, DEM aspect, and precipitation. First, the catchment area of the recharge zones must be calculated in a process similar to the watershed analysis, returning the area associated with the catchment coded as "1". A simple multiplication overlay with the precipitation data and the catchment areas yields a result giving the amount of precipitation for each catchment cell. Before using the precipitation data, we converted it to meters so as to keep units consistent. The precipitation amounts within the catchment area (*i.e.* each cell value) are to be summed, yielding the amount of water available for influx into the groundwater system. Finally precipitation per year data can be acquired.

Identifying surface areas of groundwater discharge is based on the observation that such areas provide a relatively constant source of moisture, which is readily exploited by vegetation in semi-arid region. Thus areas with year-round green vegetation are assumed to be indicators of discharge. The relative amount of vegetation in a scene can be estimated by calculating a vegetation index (Reddy *et al,* 1989 and Tucker 1979).

Essential tools for Hydrogeomorphological Mapping

Considering the complexity and heterogeneous occurrence of groundwater in different types of rocks (soft or hard rocks), it is essential to have micro-level surveys on village basis to provide the detailed groundwater information for locating the sites of wells and for recommendation of dimension and yields of wells. The areas for systematic hydrogeomorphological mapping is to be selected on certain priorities as under:

a) **Critical areas:** Areas where due to over-exploitation the natural resources like water, soil, forest, *etc* are depleting. It requires the detailed micro-level survey and delineation of critical areas for planning suitable measures for improvement.

b) **Problem areas:** The areas having problems like (i) waterlogging, (ii) water quality such as salinity, pollution *etc.* must have been separately selected and evaluated with reference to suitable measures for improvement.

c) **Potential areas:** The areas where prolific groundwater is available for exploitation such as alluvial areas, local alluvium, buried channels and watersheds in storage zones must be mapped and suggested the practices of judicious management of the water.

During hydrogeomorphological mapping it is essential to collect the historical data available through village wise maps, data sources including books, journals, tables and folders of the region. The data of well inventory, aquifer performance test and mappable hydrological units must be collected through fieldwork. In addition, the satellite imagery data would be useful in understanding physiography, geology, drainage, lineament and surface water bodies. This data should be transferred on the working map. Further, groundwater assessment data, landuse pattern and irrigation project information should be collected. After considering the above mentioned details the base map of the area under study must be prepared.

Recording of field observation requires the collection of information such as:

(a) Surface morphology,
(b) Drainage characteristics,
(c) Type of soil,
(d) Surface geology including type of formation, extent of weathering, joints and other structures,
(e) Subsurface geology *i.e.* stratigraphic sequence,
(f) Hydrological information such as aquifer type, depth of watertable, recharge conditions, hydraulic continuity and the well inventory data.

While carrying out these field observations the instruments or field kits useful are: (a) Clinometer compass or Bronton compass; (b) Dumpy level and staves; (c) Measuring tape; (d) Protractor; (e) Water level recorder; (f) Thermometer; (g) Digital portable pH meter; (h) Trajectory sticks; (i) Geological hammer and haver-sacks; (j) Toposheet maps, village maps, field note book, pencil and eraser.

Guidelines for Hydrogeomorphological Mapping/Cartography

The hydrogeomorphological map is to be prepared on 1:50,000 scales (Sheet wise). This map will contain various geomorphic units and forms. An extensive list of geomorphological units or legend is to be enclosed. Legend of the map should consist of geomorphic units, forms, structure, description and groundwater prospects (qualitative and quantitative *i.e.* liters/second). Legend for each map should be specific for the map sheet.

The ground truth should be undertaken based on pre-determined traverse plan and the doubtful units encountered during pre-field interpretation should be verified and correctly incorporated on the map. The quantitative parameters against each geomorphic unit should be indicated. The data available with the State Groundwater Department at district level can also be referred or this purpose.

The important structural trends, faults, fractures and lineaments should be depicted on the map.

A sample mock map provided, should be used as a reference or

(i) Drawing polygons (with 0.2 rotoring pen) or digitization

(ii) Transferring of hydrogeomorphological features with base map details (with 0.2 mm rotoring pen).

(iii) The details of hydrogeomorphological maps to be prepared are as follows:

a) Map (*i.e.* Map-1) showing Polygons and line features with different code numbers and 12 boundary tic points (including 4 corner tic points) on tracing film (Garware 75 microns).

b) Map (*i.e.* Map-2) showing Polygons with base map details containing legend, code, annotation, *etc.* on tracing film (Garware 75 microns). All mappable units should contain the specified codes.

(iv) One coloured copy of final map as per colour scheme given below:

Table 11.1: Colour scheme for groundwater potential.

Sr. No.	Groundwater Potential	Colour Scheme
1	Excellent and very good	Green
2	Very good to good and good	Blue
3	Good to moderate and moderate	Pink
4	Moderate to poor	Red
5	Poor to Nil and poor	Brown

Hydrogeomorphological mapping incorporates relationship of geomorphic units with their groundwater potential as interpreted from the landform characteristics (nature of landform occurrence, lithology, structure, inter-relationship with other units *etc.*) as well as sub-surface geology. The Hydrogeomorphological interpretation of the area under investigated is to be carried out on 1:50,000 scale Survey of India topographical sheet with the help of Indian Remote Sensing (IRS) False Colour Composite (FCC) imageries. Based on image characteristics of landforms (tone, texture, drainage, size, association *etc.*), their genesis, occurrence and composition, the hydrogeomorphic units of the area can be classified under fluvial origin, denudational origin and structural origin.

For the task of hydrogeomorphological mapping on 1:50,000 scale, IRS-1C, LISS III imageries are to be used adopting visual interpretation technique for delineating various geomorphic features/landform, their associated structural features, broad lithology and soil type, the thickness of weathered mantle, type of land use/land cover and for arriving at the groundwater prospects. Existing geological maps and other collateral data are also to be used, supplemented by field checks.

The maps depicting spatial information on various hydrogeomorphological units including structural lineaments and showing different groundwater prospects would form an important data base in scientific source finding approach for groundwater development plans of the specific area in particular.

Generation of thematic maps for Hydrogeomorphology

Drainage Map

Drainage information for this map has been derived from SOI toposheet and Satellite data. The drainage pattern present in SOI topographic sheet has to be digitized and drainage lines are to be superimposed on Satellite imagery digital data. Digitized drainage pattern has to be compared with satellite observed drainage pattern and a corrected drainage map is finally prepared. This map is useful for site location for harvesting of surface water and for prioritizing the watershed development.

The linking of the geomorphological parameters with the hydrological characteristics of the basin provides a simple way to understand the hydrologic behaviour of the different basins particularly of the ungauged basins. Before taking up the studies related with hydrologic simulations using the geomorphologic characteristics, the important geomorphological properties have to be quantified from the available topographical map of the basin. The geomorphological properties, which are important from the hydrological studies point of view, include the linear, aerial and relief aspect of the watersheds. Detailed definitions and description of various geomorphological parameters are described in *Chapter 7.*

The mapping of drainage pattern can be carried out using satellite data. Computation of the parameters required for morphometric analysis using manual methods like area measurement using dot grid method or using planimeter and length measurement using curvimeter are very tedious and time consuming. It is more difficult if the map is on higher scale like 1:50,000 and 1:25,000. The ordering, lengths, area and perimeter *etc.* can be easily estimated using Geographic Information System (GIS) technique. Use of GIS cannot only make this task relatively easy but accurate as well.

Landuse Mapping

Landuse map can be prepared from digital analysis of satellite data using Earth Resources Data Analysis System (ERDAS). For generation of landuse map of the area, Satellite data and PAN data have to be used. The merged scene can be re-sampled at 6-metre pixel resolution to get multi-spectral information at 6-metre resolution. Various classes of landuse *i.e.* agriculture, fallow, irrigated agriculture, forest, shrubs, open scrub, shrubs and wasteland, *etc.* can be identified in the watershed/basin area.

Soil Mapping

Soil map can be derived from supervised classification of the satellite data in conjunction with limited field data collected from field visits and general information collected from the villagers and the block office. Most of the agricultural soils usually lose the top horizon either due to erosion or construction of terraces. In the terraced hillside, the down-slope drift of mineral matter is sharply reduced and the soil is stabilized. On steep slopes, soils are generally shallow and usually have a thin surface horizon and medium to coarse texture. Sub-soils are deep and heavily textured. Top surface horizon with a high content of organic matter is a characteristic feature of the hilly area. These soils are generally highly leached and acidic in nature. Valley soils are developed from the colluvium brought down from the up-slopes. Soils of the valley bottom on river terraces comprise of alluvium, brought and deposited by rivers in the process of aggradation. All these features of soils must be marked in the soil map for perfection.

Digital Map:

The contour map and spot height map of the area are to be merged together and a composite map having information about contours as well as spot height was formed. This combined map is further interpolated at 6-metre pixel resolution using map interpolation function available in Integrated Land and Water Information System (ILWIS) to generate a digital map of the area. This digital map is further checked for flats and pits present in it. Since the area lies in steep mountainous terrain, only few flats and pits can be observed in the digital map. These flats and pits were then removed using iterative map calculation functions of ILWIS and final digital map can be generated. Removal of flats and pits in a digital map is necessary to maintain continuity of water to the catchment outlet from any point inside the catchment. This digital map is then used to delineate watershed boundary using eight-direction pour point algorithm. The digital map can be further analysed to generate slope map of the watershed. The slope map is helpful in prioritizing areas for development measures like engineering, land suitability *etc*.

The mapping procedures delineating the appropriate characteristics of the groundwater system involve three techniques such as spatial operations within a layer, reclassification of a layer, and operations between data layers. Spatial operations include both the analysis of watershed boundaries from the digital map, and the calculation of local variance in the satellite data. Reclassification of a single layer allows the analyst to group some measurement into values that meet predetermined requirements, and those that do not. In this exercise we can employ the technique primarily to perform a binary classification of "true," which is given a digital value of 1, or "false," with a digital value of 0. Operations between layers allow an analysis which links a number of data layers with the characteristic feature we wish to identify through logical assumptions or arithmetic operations. For example, overlay multiplication creates a logical decision "if both are true than the result is true" or "if one or both layers are false then the result is false." In other words, only if both data layers have overlapping cells coded "1" would the result be "1" after multiplication.

Chapter 12

Hydrogeomorphology and Social Environment

⇒ The present chapter deals with the Social – Cultural – Administrative consideration besides the education and training in hydrogeology for professionals, technicians, farmers, women and masses.

The world is heading towards a water crisis. Africa and West Asia are likely to be worst affected by water scarcity, but with increasing population, water may fall short even in the other parts of the world. According to the recent UN report, the supply of clean and fresh water is depleting at such an alarming rate in some regions, that within 30 years, about two third of population will suffer moderate to severe water stress.

In hydrogeomorphological context of social environment the water is an essential natural resource for sustaining life. The available water resources are under tremendous pressure due to increased demands. The time is not far when water, which is the free gift of nature, will become a scare commodity. Therefore, conservation and preservation of water resources are urgently required to be done at priority basis. Water management has always been practiced in our communities since ancient times. But such practices have faded away during the past few decades mainly due to lack awareness. Hydrogeologists feel that the water harvesting practices must be urgently implemented to overcome the problem of water crises. In India the Ministry of water Resources is endeavouring to make rainwater harvesting a part of every day life in villages and cities as a people's movement. Hydrogeomorphological aspect of social environment include (a) Social – Cultural – Administrative Consideration (b) Education and Training in

Hydrogeology for Professionals and Technicians (c) Education for Farmers and Women (d) Education for Masses. These are described as below:

Social – Cultural – Administrative Consideration

Management of hydrogeomorphic units including the watershed management schemes influence the social, political and cultural life of the area and it is important to consider these in the planning stage. These management activities that affect the social life include the building means of communication, such as roads, bridges, hospitals, parks, recreation centers and drainage systems. As a result the tread and commerce grow, interaction with outside world grow and there are more opportunities for social intermingling. This opening up may also accelerate migration to and from the area. It also induces cultural exchange – new ideas seep in and local culture is exposed to larger audience.

Since the activities of hydrogeomorphic units involve work on land, administrative complications due to the ownership of land, lease, rentals, *etc.,* arise and these issues might result in court cases. Unfortunately, court cases in India take a long time to settle; the time horizon may be of the order of years. All this implies cost over runs, delay and tension. The major factors determining the cost of a project are the size and areal extent, *e.g.* area for soil conservation, slope stabilization, river training, quality of water that must be stored, treated, supplied and quality of waste water to be collected, treated and disposed.

Although India is home to about 18% of the world's human population and 15% of the livestock population, it has only 2% of the world's geographical area and 0.5% of pasturelands. The per capita availability of forest in India is about 1/10th of the world average. This shows that the land, water and botanical resources are under tremendous pressure. Hence it is essential to have the better management of hydrogeomorphological resources in general and watersheds in particular. Human are always the central for concern in hydrogeomorphological context and have different roles and responsibilities as the facilitators of development, the benefactors, the beneficiaries and in some cases they have also caused the destruction of resources. The key to a better management of hydrogeomorphological resources is to apply the best available technology and organize, motivate and involve people to attain the long-term objectives of social growth and development.

Education and Training in Hydrogeology for Professionals and Technicians

In view of seriousness of problems related with hydrogeomorphological resources there is necessity of education and training for everybody associated with it. UNESCO (2000) has duly recognised this aspect while formulating the sixth phase of International Hydrology Programme (IHP – VI, 2002-2007). In the line with the comprehension of water interactions, technological development of data acquisition and improved modeling of processes and interactions, the IHP – VI topics on hydrologic research, water resources management and education are framed under five themes. One of these themes is the water education and training (WET).

According to UNESCO, it is envisaged that each of the themes within IHP – VI will aim at developing an output relevant to human capacity building by strengthening its result with training and awareness-rising activities. While WET activities are going to cover all levels aspects of education, information transfer and training, priority has been given to higher education, including institutional capacity building and networking, education for research at post-graduation level, continuing professional education and to activities targeting 'training of trainers', thus facilitating the multiplicative effect of IHP – VI efforts within the domain of education and training.

Traditionally, water resource management in India has been the domain of engineers drawn from civil and agricultural engineering departments. Now a day the researchers from physics, chemistry, geology and mathematics are also entering in this field. There is a vast difference in the curriculum of engineering and other disciplines. The engineering course on water resources cover large part in syllabus including water physics, chemistry, soil mechanics, mathematics, numerical analysis and computers while other disciplines may cover 15-20% of water related courses during the study period. Hence these students of non-professional courses need the through education and training to become expertise in the water resources field.

Studies have shown that a mere recognition of hydrological processes and phenomena are not enough to solve the complex task of society (Singh and Jain 2002). In addition to the qualitative connections of the phenomena, their quantitative characteristics should also be known; the characteristics elements of the processes, the range of their changes, and the inter-relationship between the elements should be revealed (Gilbrich 1992). Regarding the training of the serving professionals, Char and Suryanarayana (1989) recommended three-stage training such as introduction at entry, in service for intermediate level and appreciation management at senior level officer. The introduction training is aimed to help fresh graduates who enter service and it effectively prepare them to job that they are expected to perform. In the view of rapid advances in technology, the in service people require frequent trainings. As in-service people have many official and personal commitments, so they must be given short-term duration courses for training. A manager of water resources also requires the knowledge of ecology, social sciences and economics to efficiently perform his duties. Since he may have to frequently deal with public, good communication and negotiation skills are always helpful. Broad ideas of the latest tools with their strengths and weaknesses should be given to senior officers so that they accept and encourage change and improvement rather than stick with the irrelevant practices.

Another important training is the training the technicians. Technicians here means the group of people who have a supporting role in water resources management, such as observing data in the field, data entry in computers, preliminary processing and routine analysis. These people must be very well capable and should perform duties with better skills because the hydrologic design and management is based on the data that are observed by them or the analysis performed by them. Besides, the peoples in this cadre needs to be motivated from time to time so that they perform their job with devotions. The training for these people should be highly focused and the emphasis should be more on hands on training rather than academic verbal lectures. Frequently such trainings must carried out in local areas to be best of their category.

Professional organizations such as American Society of Civil Engineers (ASCE), American Institute of Hydrology (AIH) in USA,

Indian Water Resources Society (IWRS), Indian Association of Hydrologists (IAH) in India, National Institute of Hydrology (NIH) India, etc., have important role in hydrology education. These can help by organising short-term courses and public awareness campaigns. Generally, the people from professional, academic, research institutions or consultancy firm are taking part in the training activity. A large infrastructure has been created in India to meet training needs, in addition to other responsibilities. Many states have Water and Land Management Institutes (WALMIs), and engineering staff colleges. Research and training institutes of many states, such as Irrigation Research Institute (UP), Andhra Pradesh Engineering Research Institute (AP), Maharashtra Engineering Research Institute (MERI) Maharashtra, Gujrat Engineering Research Institute (GERI) Gujrat etc. are well known all over country. Research institutes like NIH and institutes under ICAR, frequently conduct short term training workshops. In addition, there are specialized institutions for training such as National Water Academy of Central Water Commission (CWC) and Groundwater Training Institute of Central Groundwater Board in India. Thus, there is a large infrastructure for education and training in the water sector in India.

Education for Farmers and Women

Farmers and women use large quantities of water and other natural resources of the hydrogeological units such as watershed and hence, they need to be brought to the central stage in management and conservation of these resources. Education for this category does not necessary involves a classroom setting. Rather, it can be best delivered in informal sessions with emphasis on how-to-do and question-answer session. Participants of this group need to be convinced of the necessity of conservation and optimum utilization of resources. The medium of communication in these sessions has to be the local language and the use of technical jargon should be minimized.

Education for Masses

Education for masses should begin with schools. Children must be given the basic information about the hydrologic cycle, water conservation, soil erosion, forestry, *etc.*, in elementary and middle school classes. Thus, an average student must know the basic things about natural resources conservation. The occasions like water resources day need to be effectively used to convey the importance, necessity and practice of watershed conservation to the public. These could be in the form of handouts, of course in local language with "dos and don'ts" and adequate illustrations. As a result, the student and public get disenchanted towards education and campaigns such as watershed conservation. The leaders, government officers and role models have to necessarily practice the things that are taught so that the good habits are reinforced and propagated.

The training of mass also must include the training for the rainwater harvesting through rooftops in urban areas. In recent times, many of our metropolitan cities have been experiencing acute shortage of drinking water especially during summer months. The shortage of ground water is more pronounced due to urbanization and limited open areas available for recharge of ground water. In some of such high demand Metropolitan areas, ground water extraction has reached a very high stage and has brought problems like declining water levels, failures of wells/tube wells, deterioration in ground water quality, saline water for drinking due to expansion of cities and increase in population. To overcome this problem it is essential to conserve every drop of rainwater that falls in such cities. Till now the rainwater has been allowed to go waste channelising it into surface streams. Time has come to look into the potentials of storing the rainwater into the underlying sub-surface porous formations to meet the ever-increasing needs of the Metropolitan cities.

As an example to quote Delhi receives around 600 mm rainfall during monsoons but the recharge to sub-surface is very less since most of the area is occupied by buildings, concrete floors, roads etc. Entire roof top rainwater is channelised into river Yamuna causing chocking of drainage system and adding flood fury of the river. Therefore, there is an urgent need to conserve most of the rainwater during monsoons for further use during lean period. Roof top rainwater harvesting for recharging ground water reservoir is a feasible proposition in urban areas. In capital territory, the roof top area works out to be about 138 sq. km, which would yield about 67.9 mcm water. Considering only the roof top rain water harvesting in 10% of area, mainly the institutional buildings in the first phase, water available for the recharge would be about 6 mcm. The schemes of roof top rainwater harvesting can improve the groundwater of the urban areas.

The education of masses must also be including with the water literacy campaign. In this context the water literacy is discussed with reference to the consumers of rural, urban and industrial sectors. Firstly, the rural consumer who can be a cultivator is to educate regarding the cropping pattern in the particular season. It is observed that actually 3% of the total area of irrigation project (reservoir) should be brought under annual large water needing crops like sugarcane and banana. The water is managed through illegal pumping of water from reservoir or canals. In view of water availability and water resource potentiality, the farmers must be taught to change the cropping pattern from intensive water needing crops to other profitable crops, which requires less quantity of water. The next group of consumer is the urban consumer. Urban consumers must know that his careless consumption of water leads to water scarcity and the Municipal body would not be in a position to supply water. Municipal councils or corporations must have implemented the water awareness campaigns. Actually, now a day the water meant for irrigation is given as water supply in the urban areas. Hence urban consumers should be made to learn that he is consuming the water meant for agriculture and he should make substantial sacrifice. The big user of water resources next to urban is the industrial utilization of irrigation water. The growing demand of industries in India is creating the pressure on the water resources. These needs are fulfilled from existing water reservoirs, hence the training is essential for optimum utilization of water.

The training and education in rural area needs to be very well documented. The village must be the central concern for water harvesting techniques. Training must be given to villagers to conserve each and every drop of the water. Starting from village every stream and hill slope must be developed to store the water as groundwater. If the villages are made self sufficient for water and natural resources then the stage will come when the villagers will mot move towards the cities. The education of water conservation will definitely enhance the water storage for future use. It must be told to villagers during education campaign that "Water is not a price less commodity, it is the elixir of life, save it, conserve it and stop the over exploitation" then and then our aim would be satisfied.

Providing safe drinking water is not possible, unless and until every user is properly educated for it's health aspects. All attempts by Government in third world fail for want of health education, therefore it has become necessary to couple drinking water program with health education program. Hence realizing that this has to be an on going activity. Till the time, entire population follows the hygienic practices as their second nature. It is proposed to undertake a programme with objectives as awareness generation for safe drinking water and health. Launch a programme for domestic chlorination and training environmental sanitation to properly manage the wastewater. The programme will be launce in 5 or 6 selected villages preferably through local women or young groups. Construction of village waste water gutters, soak pits *etc* will be the part of the programme along with, exhibitions, street plays, lectures, poster shows *etc*. Such concentrated effort over a period of year is expected to give a result that will become pace setters for other organization and villages.

References

Acreman, M. C. and Sinclair, C. D. (1986). Classification of drainage basins according to their physical characteristics; an application for flood frequency analysis in Scotland. *Journal of Hydrology*, **84**:365-380.

Adyalkar, P.G., Ayyangar, R.S., Tikekar, S.S. and Khare, Y.D. (1996). Groundwater potential of Deccan Flood Basalt of Nagpur District in Maharashtra: An imprint derived from satellite imagery In Deccan basalt, Gondwana Geol. Soc. Sp. **2**: 485-492.

Agashe, R.M. (1990). Scope for artificial groundwater recharge in Deccan trap areas of Maharashtra – an over view. Proc. *All India Seminar on modern Techniques of Rain water harvesting, water conservation and artificial recharge for Drinking water, afforestation horticulture and Agriculture.* Pune, G.S.D.A. Publ., 121-192.

Agrwal, P.K. (1987). Groundwater resources and development potential of Parbhani district, Maharashtra CGWB Rep. No. 353/DR/9/87, 1-19.

Agrwal, P.K. (1995). The Techno-economic aspect of groundwater potential and development in Deccan flood basalts, Maharashtra, *Gondwana Geol. Mag.* **10**, 79-88.

Akili, W. and Fletcher, A.H. (1978). Groundwater conditions for housing condition in the Dahran region, Eastern Province, Saudi Arabia. *Proc., International Association of Housing Science Conference,* **2**: 532-546.

Anderson,A.L.(1941). Physiographic divisions of the Colambia plateau in Idoho. J. *Geomorphology*, **4** (3), 210-222.

Anderson, M. G. and Burt, T. P. (1978a). The role of topography in controlling through flow generation. *Earth Surface Processes*, **3**:331-344.

Anderson, M. G and Burt, T. P. (1978b). Toward more detailed field monitoring of variable source areas. *Water Resources Research*, **14**:1123-1131.

Anderson, M. G. and Burt, T. P. (1990). Process studies in hillslope hydrology: an overview. Anderson, M. G. and Burt, T. P., editors, *Process studies in hillslope hydrology*, pages 1-8. Wiley.

Anonym (2000). Hydrological Cycle. Downloaded from World Day for Water 2002 web: Hydrological cycle. htm.

APHA (1989). Standard Methods for the examination of water and waste water, 17th Ed. American Public Health Association, Washington D.C., pp. 1131-1138.

Arumugam, K. and Agrawal P.K.(1990). Role of subsurface geology and physiography on yield characteristics of wells – A case study. *Proc. Vol. Of seminar on "Modern techniques of rainwater harvesting, water conservation and artificial recharge for drinking water, afforestation, horticulture and agriculture"*, pp. 642-646.

Asthana, D.K. and Asthana, M. (2003). *Environment: Problems and Solutions.* S. Chand and Co. Ltd., pp. 63-72.

Arya, A.S., Bhandari, R.J., Pathan, S.K., Patel, A.V. and Patel, S.S. (2002). Remote sensing and GIS for micro-watershed development: A grass root level approach. In proceeding volume of the international symposium of ISPRS Commission VII on Resource and Environmental Monitoring held during December 3-6, 2002, **XXXIV** *Part 7*, pp. 671-674.

Athavale, R.M., Rangrajan, R. and Muralidharan, D. (1992). Management of natural recharge in India. *J. Geol. Soc. of India,* **39**(3): 235-244.

Auren, H.F. (1991). Voronoi Diagrams – A Survey of a Fundamental Geometric Data Structure, *ACM Computing Surveys*, **23**: 3 September 1991.

Babar Md (1997). Management of environmental deterioration of pedogeomorphic surfaces in the Purna river basin, Parbhani district (Mahararashtra). *Indian J. of Geomorphology,* **2** (2), 149-158.

Babar, Md. (1999). Management of ground water resources through watershed development in Parbhani district (Maharashtra). In seminar volume on *'Common property resources and planned development in Western India challenges and prospects'* at S.R.T.M.,University, Nanded on January, 19-21, 1999, pp. 246-256.

Babar Md (2001). Hydrogeomorphological studies by remote sensing application in Akoli Watershed (Jintur), Parbhani Dist., Maharashtra, India. In *'Spatial Information Technology: Remote Sensing and GIS-ICORG'* edited by I.V. Murali Krishna, **2**: 137-143.

Babar, Md (2002a). Hydrogeomorphological mapping for groundwater resource development in the northern part of Parbhani District (Maharashtra) using IRS 1B, LISS II Geocoded data. In proceeding volume of National conference on GIS and Their Application in Civil Engineering during February 14-16 2002 at Deccan College of Engineering and Technology, Hyderabad, pp.75-83.

Babar, Md. (2002b). Application of Remote Sensing in Hydrogeomorphological Studies of Purna River Basin in Parbhani District, Maharashtra, India. In proceeding volume of the international symposium of ISPRS Commission VII on Resource and Environmental Monitoring held during December 3-6, 2002, **34** (7): 519-523.

Babar, Md. (2003). Geological and geomorphological mapping of Akoli watershed in Jintur Tahsil of Parbhani district, Maharashtra. *Indian J. of Geomorphology* **8** (1&2): 87-94.

Babar, Md. and Kaplay, R.D. (1998). Geomorphometric analysis of Purna river basin in Parbhani district (Maharashtra) India. *Indian J. Geomorphology* **3**(1): 29-39.

Babar, Md and Kaplay, R.D.(1999a). Geomorphic characteristics of Quaternary sediments of Purna river basin, Parbhani district, Maharashtra. *Gondwana Geol. Magz.*, Spl. **4**: 73-79.

Babar, Md and Kaplay R.D. (1999b). Groundwater quality around Pingalgad Nala, Parbhani District. *J. Ecology, Environment and Conservation,* **5**(2): 141-142.

Babar, Md. and Kaplay, R.D. (2003). Groundwater Fluctuation in Purna River Basin Parbhani District, Maharashtra. *J. of Applied Hydrology,* **16** (l): 56-6l.

Babar, Md. and Kaplay, R.D. and Panaskar, D.B. (2000). Structural control on drainage pattern in the Tirna River Catchment, Central west India. *Indian J. of Geomarphology,* **5** (112): 126-133.

Babb, J. S.; Cole C. A.; Brooks R. P., and Rose A. W. (1997). Hydrogeomorphology, Watershed Geology, and Water Quality of Wetlands in Central Pennsylvania. *J. of the Pennsylvania Academy of Science.* **71** (1): 21-28.

Bagtzoglou, A. C., Islam, M. R., and Muller, M. (1994). Stochastic analysis of unsaturated flow and transport with the SUFLAT executive numerical code. In *Proceedings of Fifth International Conference on High-Level Radioactive Waste Management*. American Nuclear Society & American Society of Civil Engineers, pp. 2669–2677.

Baker, V.R. and Kochel, R.C.(1988).Flood sedimentation in bed rock fluvial systems. In Baker, V.R., Kochel, R.C. and Patton, P.C.(eds.), *Flood geomorphology*, John Wiley and Sons., New York, pp.123-137.

Balasundaram, M.S.(1977).Contribution to geomorphology and geohydrology of the Brahmaputra valley. Geol. Surv. India Misc. Pub., No. 12.

Balasundaram M S & Rao G S M (1972), History and development of engineering geology, *India Records*, **104** (2):1-22

Band, L. W. 1986. Topographic partition of watersheds with digital elevation models, *Water Resour. Res,* **22**: 15-24.

Barrett, E. C. and Curtis, L. F. (1995). Introduction to Environmental Remote Sensing, 3rd ed. Chapman & Hall, London. 426 pp.

Beane, J.E., Turner, C.A., Hooper, P.R., Subbarao, K.V. and Walsh, J.N. (1986). Stratigraphy, composition and form of the Deccan basalts, Western Ghats, *India. Bull. Volcanol.,* **44**: 61-83.

Bedi, N. and Bhan, S.K. (1978). Application of Landsat imagery for hydrogeological mapping in Cuddapah area *A.P. Proc. of Joint Indo-UK Workshop on Remote sensing of water resources*. NRSA Hyderabad, pp 115-129.

Berg, R. R., and Haveman-Avery, A. 1995. Sealing properties of Tertiary growth faults, Texas Gulf coast. *American Association of Petroleum Geologists Bulletin* **79**(3): 375–393.

Betz F. Jr. (1975). Environmental Geology. Straoudsberg Penn., Dowden, Hutchinson and Ross.

Beven, K. (1995) Grid ATB_Sink: Program to Identify and Correct Sink Values from Gridded Elevation Data, Version 95.01. Centre for Research on Environmental Systems and Statistics, Lancaster University, Lancaster.

Beven, K. and Kirkby, M. J. (1979). A physically based, variable contributing area model of basin hydrology. *Bulletin of Hydrologic Sciences*, **24**: 43-69.

Bhagavan, S.V. BK(1998) Remote sensing techniques for watershed programmes in Andhra Pradesh. *Proc. of workshop on watershed management Problems and Prospects* (WMPP98), Centre for Water Resources, J.N. Technological Univ., Hyderabad, pp. 24-33.

Bhagavan, S.V.B.K. and Raghu, V. (1998). Implementation of remote sensing based action plan and impact analysis Anantpur Sist., Andhra Prasdesh. *Proc. of workshop on watershed Management –Problems and Prospects*

(WMPP 98), Centre for Water Resource) J.N. Technological Univ. Hyderabad pp. 34-42.

Bhan S.K. (1998) Remote sensing for national development and trends in remote sensing. *Proc. of workshop on watershed management-Problems and Prospects* (WMPP-98), Centre for Water Resources, J.N. Technological Univ, Hyderabad, pp. 14-23.

Bhatt, M.K. and Salpekar, P.R. (1990). Artificial recharge through bore wells in proble village Kaij, District Beed. Proc. Vol. Of seminar on *"Modern techniques of rainwater harvesting, water conservation and artificial recharge for drinking water, afforestation, horticulture and agriculture"*, pp. 271-274.

Bhattacharya, G (1996), Remote sensing for geological, geomorphological and ground water studies, 16th Indian Geography Congress, pp.181-303.

Bhattacharya, A and Banerjee, S.N. (1979). Quaternary geology and geomorphology of Ajay – Bhagirathi valley, Birbhum and Murshidabad district, West Bengal. *Indian J.Earth Sci.* **6**(1): 91-102.

Bhushan, Brij, (2002). Using GIS for crop forecasting and crop estimation in India. In *'Spatial Information Technology: Remote Sensing and GIS-ICORG'* edited by I.V. Murali Krishna, Vol.-I: 197-203.

BIS (1991). Indian standard specification for drinking water, BS 10500.

Black, C.A. (1965). *Methods of soil analysis. Part–2* Amer. Soci. Agron-Inc. Medison, Wisconsin, p.581.

Blanchet, P.H. (1957), *Development of fracture analyses as exploration methods: Bulletin, American Association of Petroleum Geologists*, **41**: 1748-59.

Bloom, A.L.(1978). *Geomorphology*. Printice-Hall of India Private Ltd., New Delhi, pp. 26-89.

Blöschl, G. and Sivapalan, M. (1995). Scale issues in hydrological modeling: a review. In Kalma, J. D. and Sivapalan, M., editors, *Scale issues in hydrological modeling*, 9-47. Wiley.

Bolstad P V, Stowe T (1994), *An evaluation of DEM accuracy: Elevation, slope and aspect, Photogrametric engineering and remote sensing:* 1327-32.

Braticevic, D. and Karanjac, J. (1997). Ground Water for Windows (ver. 1.31) software and manual, Department for Development Support and Management Services, United Nation.

Brinson, M. M. (1993). Hydrogeomorphic Classification for Wetlands. Final Report. Vicksburg, Miss: U.S. Army Corps of Engineers Waterways Experiment Station; WES/TR/WRP-DE-4. 103 p.

Brinson, M.M. (1996) Assessing Wetland Functions Using hydrogeomorphology. National Wetlands Newsletter; **18** (1): 10-16.

Brown, E.H. (1980). *Geography: Yesterday and Tomorrow*. Ed. Brown E.H. Oxford Univ. Press, London.

Bruner, H. (1970). Pleistoozaene Kilmaschewan Kungen in Bereich des Oestichen Mysore plateau (Sued Indian). Geologie, **19**: 72-82.

Bulusu, K.S. and H.C. Arora (1967). Certain observations on self purification of Khan river and its effect on Krishana River. *Indian J. Env. Health*, **9**: 275-295.

Burrough, P. A., (1986). *Principles of Geographical Information Systems for Land Use Assessment*. Oxford University Press, New York, pp. 194.

Burwell E B, Roberts G D (1950), The geologist in the engineering organization: application of geology to engineering practices, Berkey volume, pp. 327, G S A, New York.

Calef, W.C. (1950). Slope studies of Narther Illinois. Trans Illinos Academy of Science. **13**: 110-115.

Caran, Christopher S., Woodruff, C.M., Jr., and Thompson, Eric J., (1982), Lineament analysis and inference of geologic structure—examples from the Balcones/Ouachita Trend of Texas: University of Texas at Austin, Bureau of Economic Geology Geological Circular: **82**-1: 59-69.

Carlass, T.G. (1838). *Delta of India. Jour.* Roy. Geographical Society **8**.

Carter J R (1988), Digital representation of topographic surface: Photogrametric engineering and remote sensing.

Chapelle, F.H. 1993 *Groundwater microbiology and geochemistry.* J. Wiley and Sons, New York, pp. 424.

Char, S.A. and Suryanarayana, N. (1989). Education and training requirements in various disciplines and levels in water resource requirement up to 2025 (a) Quantitative projection of man power requirements (b) Course content. *Proceedings of International Conference on Education and Training in Water Resources*, held at Aurangabad. Published by CBIP, New Delhi.

Chavan, S.P., Joshi, P.K., Pawar, S.B. and Md. Babar (2002). Qualitative analysis of planktons in wells around Pingalgad Nala in Parbhani District (Maharashtra) India. In: *"Biotechnology in Agriculture, Industry and Environment"-Proc. Vol. Of International Conf. of SAARC Countries* (A.M. Deshmukh ed.), pp.336-342.

Cheeney R.F (1983) Statistical methods in Geology, George Allen Unwind Ltd., London, p. 169.

Chilton, P.J. and West, J.M. 1992 Aquifers as environments for microbial activity. In: *Proceedings of the International Symposium on Environmental Aspects of Pesticide Microbiology,* Sigtuna, Sweden, pp. 293-304.

Chorley, R.J. (1957). Climate and morphometry. *J.Geol.* **65**: 628-638.

Chorley, R.J. (1969). The drainage basin as the fundamental geomorphic unit. In water, Earth and man (ed. By R.J.Chorley), Methuen, London. pp. 77-100.

Chorley, R.J. (1971). The role and relations of Physical geography. *Progress in Geography*, **3**: 87-109.

Chow, V.T. (1966) *Handbook of Applied Hydrology*. McGraw Hill Book Co., New York.

Clark, L. and Baxter, K.M. 1989 Groundwater sampling techniques for organic micropollutants: UK experience. *Quart. J. Eng. Geol.,* **22**(2): 159-168.

Clarke, J.I.(1970). Morphometry from maps. *Essays in Geomorphology*. (ed. by G.H.Dury). Heinmann, London. pp. 235-274.

Clayton, K.M. (1980). *Geography: Yesterday and Tomorrow*. Ed. Brown E.H. Oxford Univ. Press, London: 167-180.

Cole, C. A.; Brooks R. P., and Wardrop D. H. (1997) Wetland Hydrology as a Function of Hydrogeomorphic (HGM) Subclass. *Wetlands*. **17** (4): 456-467.

Conacher, A.J. and Dalrymple, J.B. (1977). The nine unit land surfaces model: an approach to pedogeomorphic research. *Geoderma*, **18**.: 1-154.

Cooke, R. U. and Doornkamp, J. C. (1990). *Geomorphology in environmental management*. Oxford University Press, 2nd ed.

Corvinus, G.,Rajaguru,S.N.and Mujumdar, G.G. (1973).Some observations on the Quaternary formations of Western Maharashtra (India). *Quaternary*, **23-24**: 53-69.

Cox, K. G. and Hawkeswarth, C.J. (1984). Relative contribution of crust and mantle to flood basalt magmatism, Mahabaleshwar area, Deccan traps. Phill. Trans. R.Soc. London, **A310**: 627-641.

Crave, A. and Davy, P. (1997). Scaling relationships of channel networks at large scales: Examples from two large magnitude watersheds in Brittany, France. *Tectonophysics*, **269**: 91-111.

Crickmay, C.H. (1933). The latter stages of cycle of erosion. *Geol. Magz.*, **70**: 337-347.

Das S, Behera S C, Kar A, Narendra P and Guha S (1997). Hydrogeomorphological Mapping in ground Water Exploration using Remote Sensed Data- A case study in Keunjhar District, Orissa. *J. Indian Soc. Remote Sensing* **259**(4): 247-260.

Davis, W.N. (1899). The peneplain. In geographical essays edited by D.W. Johnson, Dover publ., INC. (1909), pp.350-380.

Davis, S.N. and De Wiest, RJM 1966 *Hydrogeology.* John Wiley, New York, pp. 463.

De, A.(1994). Fractionation trends of the Deccan trap magmas in parts of Western Gujarat and Satpuras. In: Subbarao K.V. (ed), volcanism, Wiley Eastern Ltd., pp.277-294.

Deolankar, S.B. (1980). The Deccan Basalts of Maharashtra India: their Potential as aquifers. *Groundwater* **19**(5): 434-437.

Deshmukh, S..S. (1980). Geology, Petrography and Petro-chemistry of the Deccan basalts in parts of Maharashtra state. Unpublished progress report of G.S.I., CR, Nagpur.

Devey, W.C. and Cox, K.G.(1987). Relationship between crustal contamination and crystallisation in continental flood basalt magams with special reference to the Deccan traps of the Western Ghats, *India. Earth Planet.Sci.Lett.* **84**: 59-68.

Devey, W.C. and Lightfoot, P.C.(1986). Volcanological and tectonic control of stratigraphy and structure in the Western *Deccan Traps Bull. Volcanol.*, **48**: 195-207.

Dhembare, A.J., Pondhe, G.M. and Singh, C.R. (1998). Groundwater characteristics and their significance with special reference to public health in Pravara area, *Maharashtra. Pol. Res.* **17**(1): 87-90.

Dhokrikar, B.G.(1991). Groundwater resources development in basaltic rock terrain of Maharashtra, (ed. by B.G.Dhokrikar) Water Industry Publ., Pune, pp. 122-185.

Dhruva Narayana, V. V. (1987). Downstream impacts of soil conservation in the Himalayan region, *Mountain Research and Development* **7**(3): 287-98.

Dietrich, W.E., Dunne, T., Humphrey, N.E. and Reid, L.M.(1982). Construction of sediment budgets for drainage basins. In: Sediment budgets and routing in forested drainage basins, USDA Forest Service General Tech. Rep. PNW – 141, PP. 5-23.

Dikau, R. (1989). The application of a digital relief model to landform analysis in geomorphology. In Raper, J., editor, *Three dimensional application in Geographic Information Systems*, pp. 51-77. Taylor & Francis.

Dikau, R. (1994). Computergestützte geomorphographie undihre anwendung in der regionalisierung des reliefs. *Petermanns Geographische Mitteilungen*, **138**: 99-114.

Dikau, R., Brabb, E. E., Mark, R. K., and Pike, R. J. (1995). Morphometric landform analysis of New Mexico. In *Advances in geomorphometry - Proceedings of the Walter F. Wood Memorial Symposium*, Vol. 101 of *Zeitschrift für Geomorphologie, Supplement Band*, pp. 109-126.

Dikshit, K.R. (1970).Polycyclic landscape and the surfaces of erosion in the Deccan Trap country with special reference to upland Maharashtra. *The Nat. Geog. Jour. India,* **16**: 236 – 252.

Domenico, P.A. and F.W. Schwartz, (1990). *Physical and Chemical Hydrogeology*, John Wiley & Sons, New York, 824 p.

Doornkamp, J.C. and King, C.A.M. (1971). *Numerical analysis in Geomorphology – An introduction.* St. Martins Press, New York, 372 p.

Downey, M. W. (1984). Evaluating seals for hydrocarbon accumulations. *American Association of Petroleum Geologists Bulletin* **68**(11): 1752–1763.

Driscoll, F.G. 1986 *Groundwater and Wells.* 2nd ed., Johnson Division, St Paul, Minnesota, 1089 pp.

DST - CSRE (1999). GRAM++ Window based GIS Package, Deptt of Science & Technology, New Delhi.

Dumout, J.F., Mering, C., Parrot, J.F. and Tund, H.(1996). Morphological and mathematical analysis of asymmetrical fluvial pattern: A study case from the Vcayali river (Peru). Z. Geomorph. N.F. Suppl. ed. **103**: 269-82.

Dury, G.H. (1972). Some current trends in Geomorphology. *Earth Science Review*, **18**, pp.45-72.

Eash, D. A. (1993). Estimating design-flood discharges for streams in Iowa using drainage basin and channel-geometry characteristics. United States Geological Survey Water Resources Investigations Report, **93**: 4062-96.

Endreny, T.A. Wood, E.F. and D. P. Lettenmaier (2000) Satellite-derived digital elevation model accuracy: Hydrogeomorphological analysis requirements. *Hydrol. Process,* **14**: 1-20.

Early, J. (2001). The Influence of Linear Features on Groundwater Flow in the Fractured Carbonate Aquifer in Berkeley and Jefferson Counties, West Virginia, 493-590.

Easterbrook, D.J.(1969). *Principles of geomorphology*. Ed. by D.J.Easterbrook, McGraw Hill; New York, pp. 1-46.

Ehrlich, H.L. (1990) *Geomicrobiology.* 2nd Ed. Marcel Dekker, New York, pp. 646.

Ernst, L. F. and N. A. de Ridder (1960). "High resistance to horizontal ground-water flow in coarse sediments due to faulting." *Geologie en Mijnbouw* **39**: 66-85.

Evans, J. P., C. B. Forster and G. J.V. (1997). "Permeability of fault-related rocks, and implications for hydraulic structure of fault zones." *J. of Structural Geology* **19**: 1393-1404.

Eyles, R.J.(1965). Slopes studies on the Wellington, Peninsular New Zealand. *Geographer*, **21**: 133- 143.

Fairbridge, R.W. (1968). *Encyclopedia of Geomorphology*, Reinhold, New York, pp. 1295.

Fairfield, J. and Leymarie, P. 1991. Drainage networks from grid digital elevation models, *Wat. Resour. Res.,* **27**: 709-17.

Fennaman, N.M.(1914). Physiographic boundaries within the United States, *Ann.Asson.Amer.Geogr.* **4**: 84-134.

Fergussan, J.(1863).Delta of Ganges. *Quart. Jour. Geol. Soc.,* **19**: 321-354.

Fetter, C.W. (1994). *Applied Hydrogeology*, 3rd ed. Upper Saddle River, NJ: Prentice Hall, Inc.

Foster, S.S.D., Ventura, M. and Hirata, R. 1987 *Groundwater Pollution: An Executive Overview of the Latin America-Caribbean Situation in Relation to Potable Water Supply.* Pan American Centre for Sanitary Engineering and Environmental Sciences, Lima, pp. 38.

Freeze, R. A. and Cherry, J.A. (1979). Groundwater. Prentice-Hall Inc., New Jersey, 608 pp.

Gaile G L and Burt J A (1980). Directional Statistics, Concept and techniques in modern geography, No.25. *Geo Abstracts*, Univ.East Anglia, Nkorwish, 39 pp.

Gardiner, V.(1980). Drainage basin morphometry: Quantitative analysis of drainage basin form. In: H.S.Sharma (eds.) *Perspective in Geomorphology,* **2**: 107 –142.

Gerrard, J.(1992). Soil geomorphology: An integration of pedology and geomorphology, (ed. by J.Gerrard) Chapman and Hall, pp. 1-9.

Ghose R (1993). Remote sensing for analysis of groundwater availability in an area with long unplanned mining history. *J.Indian Soc. Remote Sensing,* **21**(3): 119-126.

Ghosh, R.N. and Mazumder, S.(1981).Neogene-Quaternary sequence of Kasai basin, West Bengal, Indian Proc. Field Conf. Neogene-Quaternary boundary, India (ed. by Sastri *et al.,* 1981) Geol. Surv. India, pp. 63-73.

Gilbrich, W.H. (1992). Some remarks on UNESCO's program in hydrological education under IDP/IHP. In Hydrology AND water Resources Education, Training and Management, edited by J.A. Raynal. Water Resources Pub. Colorado.

Godbole, S.M., Rana, R.S. and Natu, S.R. (1996). Lava stratigraphy of Deccan basalts of Western Maharashtra. *Gondwana Geo.Mag.,* Spl. **2**., 125-134.

Golterman,H.L., Clymo, R.S. and Ohnstad, M.A.M. 1978 Methods for Physical and Chemical Analysis of Freshwater. IBP Hand Book No.8, Blackwell Scientific Publications.

Goswami, A.B.(1993). Review on Quaternary Geological Mapping and standardisation of scheme of classification for inter basin regional correlation of Quaternary landform and Geological units in East and North-East *India. Geol. Surv. India,* Rec. **121**, (Pt. 2-8), 55-85.

Goswami, D.C., Goswami, I.D., Duarah, B.P. and Deka, P.P.(1996). Geomorphological mapping of Assam using satellite remote sensing technique. *Indian J. Geomorph.* **1** (2), 225-235.

Gravelius, H.(1914). Flusskunde, Band 1, Goschenesche Verlagshandlurg, Berlin and Leipzing, P. 176 (Quoted from Horton, 1945).

Gregory, K. J. and Walling, D. E. (1973). *Drainage basin. Form and Process*. Edward Arnold.

Gumbel E. J Greenwood J. A and Durand D (1953). The circular normal distribution: Tables and theory. *J. Amer.Soc.,* **18**, 131-152.

Gupta, R.D., Garg, P.K. and Arora, M.K. (2002). A GIS based spatial modeling for developmental planning. In *'Spatial Information Technology: Remote Sensing and GIS- ICORG'* ed. by I.V. Murali Krishna, **1**: 265-271.

Gupta, V. A., Waymire, E., and Wang, C. T. (1980). A representation of an instantaneous unit hydrograph from geomorphology. *Water Resources Research*, **16**(5):855-862.

Gupta, V. K., Rodruigez-Iturbe, I., and Wood, E. F., editors (1986). *Scale Problems in Hydrology*. D. Reidel.

Gupte, R.B. (1971). Engineering aspects of Deccan Trap Basalt, *Maharashtra Public Work Journal* **7**(3-4), 7-20.

Gyasi-Agyei, Y., Willgoose, G. and DeTroch, F. P. (1995). Effects of vertical resolution and map scale of digital elevation models on geomorphological parameters used in hydrology, *Hydrol. Process.,* **9**: 363-382.

Hammond, E.H.(1954). Small scale continental landform maps, *Ann. Asson. Amer. Geogr.* **XLIV**, 33-42.

Hammond, E.H.(1965).What is landform? Some further comments, *The Professional Geogr.*, **XVII**, No.3, 12-14.

Hammond, E. H. (1964). Analysis of properties in landform geography: an application to broad scale landform mapping. *Annals of the Association of American Geographers*, **54**:11-19.

Hare, F.K. (1973). Energy based climatology and its frontier with ecology. In Chorley, R.J. (ed.) Dictionaries in Geography, London Methuen, Chapter 8.

Hattingh, J.(1996). Fluvial response to allocyclic influences during the development of the lower Sundays river, Eastern Cape, *South Africa. Quaternary Intl.,* **33**: 3-10.

Hedberg, H.D. (1976). International stratigraphic guide to stratigraphic classification, terminology and procedures. John Wiley and Sons (Quoted from Godbole *et al,* 1996).

Helmlinger KR, Kumar P, Foufoula-Georgiou E. 1993. On the use of digital elevation model data for Hortonian and fractal analyses of channel networks. *Water Resources Research,* **29**(8), 2599-2613.

Helpke, C. (1995). State-of-the-art of digital photogrammetric workstations for topographic applications. *Photogramm. Engg Rem. Sens.,* **61**: 49-56.

Hem, J.D. 1989 *Study and Interpretation of the Chemical Characteristics of Natural Water.* Water Supply Paper 2254, 3rd edition, US Geological Survey, Washington, D.C., pp. 263.

Hendrickx, J. M. H. (1990). Determination of hydraulic soil properties. In Anderson, M. G. and Burt, T. P., editors, *Process studies in hillslope hydrology*, pp. 43-92. Wiley.

Hennrich, K., Schmidt, J., and Dikau, R. (in press). Regionalization of geomorphometric parameters in hydrologic modelling using GIS. IAHS Publications.

Heywood, I, Corneluis S, Carver, S. (2003): *An Introduction to Geographic Information Systems*, Pearson Education, 1st Indian Ed., 295 p.

Horton, R.E.(1932). Drainage basin characteristics.Trans. Amer. Geophys. Un., pp. 350-361.

Horton, R.E.(1945). Erosional development of streams and their drainage basins: Hydrophysical approach to Quantitative morphology. *Bull. Geol. Soc. Am.* **56**, 275-370.

Howard, A. D. (1967). Drainage analysis in Geologic Interpretation: A summation. *The Amer. Assoc. Petroleum Geologists. Bull.* **51** (1), 2246-2259.

Howrd, A.D. and Spock, L.E.(1940). Classification of landforms. *J.of Geomorphology.* **3**(4), 332-354.

Huberth, P. M. (1994). A Hydrogeomorphic Classification Model for Forested Wetlands of Southeast Alaska. Managing Forests to Meet People's Needs. *Proc. of the Society of American Foresters;* Anchorage, AK. 1994: p. 313-319.

Hukku B M (1975), Bhakra Nangal project, Punjab, G S I, Engineering Geology case studies, Misc Pub. Nos. **29**: 11 - 13.

Hussainy, S.U. (1967). Studies on limnology and primary production of a tropical lake. *Hydrobiologia*, **30**: 335-352.

ICMR (1975). Manual of standards of quality for drinking water supplies. Indian Council of Medical Research, Sep. Rep. Ser. No. **44**: 27.

IMSD (1995). Integrated mission for sustainable development: Technical guidelines. National Remote Sensing Agency, Hyderabad, Publ., pp. 11-21.

Iqbaluddin, SaifUddin and Akram Javed (1997).Geomorphology and landscape evolution of Bharatpur district, Rajasthan. *J. Indian Soc. Remote Sensing,* **25**(3): 177-186.

ISI (1983) Indian standard specification for drinking water ISI 10500.

Jain, K. and Reddy, N. (2002). Map generalization using GIS. In *'Spatial Information Technology: Remote Sensing and GIS- ICORG'* ed. by I.V. Murali Krishna, Vol. **1**: 254-260.

Jain P.K. (1996). Hydrogeochemistry and ground water quality of Singhari river basin, dist. Chhatarpur (MP). Poll. Res. **15**(4): 407-409.

Jensen, J. R. 1996. *Introductory Digital Image Processing: A Remote Sensing Perspective.* Prentice Hall, New Jersey. 316 pp.

Jenson, S. K. and Dominique, J. O. (1988). Extracting topographic structure from digital elevation data for geographic information system analysis, Photogramm. *Engg Rem. Sens.*, **54**: 1593-1600.

Jet Propulsion Laboratory (JPL) 1997. Shuttle Radar Topography Mission, JPL 400-713. National Aeronautics and Space Administration, Pasadena, California.

Joerg, W.L.G.(1914).The sub divisions of North America into natural regions: A preliminary inquiry, *Ann. Asson. Amer. Geogr*. **4**: 55-83.

Jones, D.K.C. (1976). British Applied Geomorphology: An appraisal. Zeitschrift for geomorphologie, supplement, **36**: 48-73.

Jones, D.K.C. (1983). Environments of concern. Transactions, Institute of British Geographers, **8**(4): 429-457.

Joshi A., Sharma, A.P. and Pant, M.C. (1986). Limnological studies in a subtropical system (Lake Sutlej, India). Acta Hydrochim. *Hydrobiology*, **9**(4): 407-425.

Joshi, A.K., Jayaram, A., Krishna Murthy, Y.V.N. and Srinivasan, D.S.(1996). Integrated geological study of Deccan Trap province of Maharashtra and adjoining parts. *Gondwana Geol. Magz.,* Spl. **2**: 461-66.

Kale, V.S.(1983).The Indian Peninsular movement Western Ghat formation and their geomorphic repercussion. *Trans. I.I.G.* **5** (2): 145-156.

Kale, V.S., Ely, L.L., Enzel, Y. and Baker, V.R.(1994). Geomorphic and hydrologic aspects of monsoon floods in Central India. *Geomorphology*, **10**: 157-168.

Kale, V.S. and Kulkarni, H (1992). IRS-1A and LANDSAT data in mapping Deccan Trap flows around Pune : Implications on hydrogeological modeling. Archives. *Int. Soc. Photogramm and Rem. Sens.,* **29**: 429-435.

Kandasamy, V. and Venugopal, K. (2002). Generation of resource information system and action plan for a Mynali watershed in the Nilgiri District, India using remote sensing, GPS and GIS. *In proceeding volume of the international symposium of ISPRS Commission VII on Resource and Environmental Monitoring* held during December 3-6, 2002, **XXXIV** *Part 7*: 934-937.

Karanth, K.R. (1989). *Hydrogeology*: Tata McGraw Hill Publ. Co. Ltd., New Delhi, pp. 1-455.

Karanth, K.R. and Seshu Babu, K. (1978). Identification of Major lineaments on satellite imagery and on aerial photographs for delineation of possible potential groundwater zones in Penukonda and Dharmavaram talukas of Anantpur district. *Proc. of Joint Indo US workshop on Remote sensing of water resources,* NRSA Hyderabad, pp.188-197.

Khadri.S.F.R., Subbarao, K.V., Hooper, P.R. and Walsh, J.N.(1988). Stratigraphy of Thakurwadi formation, Western Deccan basalt province. *Geol. Soc. India,* Mem. **10**: 281-304.

King, L.C.(1950).The study of the world's plain lands, *Quat. J. Geol. Soc., London,* **106**: 101-131.

Kirkby, M. J. (1988). Hill slope runoff processes and models. *J. of Hydrology*, **100**: 315-339.

Kleeberg, H.-B., editor (1992). *Regionalisierung in der Hydrologie*. Deutsche Forschungsgemeinschaft.

Klimentov, P.P (1983). General Hydrogeology. Klimentov, P.P (ed.), Mir Publishers, Moscow, pp. 86-115.

Kodarkar, M.S., Diwan, A.D., Murugan, M., Kulkarni, K.M. and Anuradha Ramesh (1988). Methodology for water analysis. Indian Association of Aquatic Biology (IAAB), Pub. No. 2

Konrad, C.E. (1998). A flood climatology of the lower Roanoke river basin in North Carolina, *Physical Geography*, **19**(1): 15-34.

Kraus, M.J. (1997). Lower Eocene alluvial paleosols: Pedogenic development, stratigraphic relationship and paleosol/landscape associations. Palaeogeography, *Palaeoclimatology and Palaeoecology,* **129**: 387-406.

Krishnaswami, V.S. (1972). Systematic geotechnical studies in the country records, *GSI*, **104** (2): 1330

Krishnaswami, V.S. (1982). History and Development of engineering geology and geotechnology in India, *Proc. Internat. Assoc. of Eng. Geol.,* **III**, 361-369.

Kulkarni H. (1992). Delineation of shallow Deccan basaltic aquifers from Maharashtra , using aerial photo interpretation. Photonirvachak. *J. Indian Soc. Rem. Sensing,* **20** (2&3): 129-138.

Kulkarni, H., Deolankar, S.B., Lalwani, A. and Lele, V.A (1994). Integrated remote sensing as an operational aid in hydrogeological studies of Deccan basalt aquifer. *Asian-pacific remote sensing J.* (ESCAP), **6**(12): 9-18.

Kulkarni, H. and Deolankar, S.B. (1995). Hydrogeological mapping in the Deccan Basalt-An appraisal *J. Geol. Soc. Indian* **46**(4): 345-352.

Kulkarni, M.N., Sunil, P.S. and Ponraj, M. (2002). Crustal deformation studies in the Western Maharashtra region of Indian Peninsular Shield using GPS. In *'Spatial Information Technology: Remote Sensing and GIS-ICORG'* edited by I.V. Murali Krishna, **1**: 315-319.

Kulkarni, S.R., Karmarkar, B.N. and Gupte, R.B. (1980) Variation in the nature of Deccan trap volcanicity of western Maharashtra in time and space. Memoir Geol. Soc. India, No.3, pp. 143-152.

Kumar Ashok and Prasad, B.B. (1993). "Correlation of lineaments with surface and bed rock topography in Lapasiya watershed of Siwane sub-basin, Hazaribagh, Bihar", *Proc. of National Symposium or Remote Sensing and Annual Convention,* Nov.22-24 Ludhiana, ISRS, pp. 166-172.

Kumar Ashok and Prasad, B.B. (2002). DSTM and DBTM for estimation of static and dynamic groundwater reserves in hard rock – A case study in Margajo watershed. *In proceeding volume of the international conference on Hydrology and Watershed Management* held during December 18-20, 2002, **1**: 369-378.

Kumar, Ashok, Prasad, L.B., Prasad, B.B. and Nisha Mendiratta (2001). GIS and GWW is tool for creating Ground Water Information System (GWIS) - a case study of upper Barakar basin, Bihar. In *'Spatial Information*

Technology: Remote Sensing and GIS- ICORG' edited by I.V. Murali Krishna, **II**: 145-151.

Kumar Ashok, Prasad B.B., Sinha Ranjan (1996). "Sustainable utilisation and management of groundwater in Churchu watershed, Hazaribagh, Bihar -A remote sensing, geophysical and GIS approach ", presented in ISRS symposium, 4- 6 Dec., Pune.

Kumar, Om and H.N. Mathur (1989). Water quality status in Doon valley forest. *Indian Forester,* **115**(8): 525-535.

Lamber, R. (1975). Recherches Hydrologiques dans le Sudest du Basin Garonnais. Doctoral Thesis, Tome, 1, Tomouse, (Quoted from Ion. Zavoianu, 1985, pp. 63-87).

Langbein, W.B. (1947). Topographic characteristics of Drainage basins, U.S.Geol. Survey water supply paper, 968, pp. 125-157.

LaRiccia, M.P. , and Rauch, H.W., (1977), Water well productivity related to photo-lineaments in carbonates of Frederick Valley, Maryland, in Dilamarter, R.R., and Csallary, S.C., eds., Hydrologic Problems in Karst Regions: Western Kentucky University, Bowling Green, Kentucky, p. 228-235.

Lattman, L.H. and Porizec, R.R.(1964). Relationship between fracture traces and occurrence of groundwater in carbonate rocks, *J. Hydrology,* **2** (2): 73-91.

Lawrence, A.R. (1985). An interpretation of dug well performance using a digital model. Groundwater, **23**(4): 449-454.

Lawson. A.C. (1932). Rain wash erosion in humid regions. *Bull. Geol. Soc. Amer.* **43**: 703-724.

Lecee, S.A. (1997). Spatial patterns of historical over bank sedimentation and floodplain evolution, Blue River, Wisconsin. *J. Geomorphology,* **18**: 265-277.

Leeder, M.R. (1997). Sedimentary basins: Tectonic recorders of sediment discharge from drainage catchments. *Earth surface Processes and Landforms*, **22**: 229-237.

Lekach, J., Amit, R., Grodek, T. and Schick, A.P. (1998). Fluvio-pedogenic processes in an ephemeral stream channel, Nahal Yeel, Southern Negav, *Israel. J. Geomorphology*, **23** (2-4): 353-369.

Lewis, W.J., Foster, S.S.D. and Drasar, B.S. (1982) *The Risk of Groundwater Pollution by On-site Sanitation in Developing Countries.* IRCWD Report 01/82, IRCWD, Duebendorf, pp. 79.

Lighfoot, P.C. and Howkesworth, C.J. (1988). Origin of Deccan trap lavas: evidences from combined trace element and Sr, Nd and Pb isotope studies. *Earth Planet. Sci. Letter*, **91**: 89-104.

Linton, D.L. (1951). The delimitation of Morphological regions London. Essays in Geography (L.D. Stamp and S.W. Wooldridge, Eds.), Univ. of London Publ., pp. 199-217.

Lohman, S.W. (1972). *Groundwater hydraulics.* U.S. Geol. Survey Prof. Paper, 708.

Lvovitch, M.I. (1972). World water balance: general report. In: IASH/ UNESCO/WMO *Proceedings of Symposium on World Water Balance,* Reading 1970, IASH Proceedings No. 2, International Association of Hydrological Sciences, Wallingford, UK, pp. 401-415.

Mahajan, G. (1993). Groundwater Recharge. Ashish Publication House, New Delhi, pp. 143.

Mahoney, J.J., McDougall, J.D., Lugmair, G.W., Murali, A.V., Sankardas, M. and Gopalan, K. (1982). Origin of the Deccan trap flows at Mahabaleshwar inferred from Nd and Sr isotopic chemical evidence. *Earth Planet. Sci. Letter,* **60**: 47-60.

Maidment, D.R. (2002). Arc Hydro – GIS for Water Resources, ESRI, 380 New York Street, Redlands, California 92373-8100.

Maji, A.K., Tamgade, D.B., Obi Reddy, G.P. and Gajbhiye, K.S. (2002). GIS based soil information system of Bilaspur District (M.P.) India. I*n proceeding volume of the international symposium of ISPRS Commission VII on Resource and Environmental Monitoring* held during December 3-6, 2002, Vol. **XXXIV** *Part 7*: 746-750.

Malewar, G.U.(1976). Placement of black soils of Marathwada in the comprehensive system of soil classification. *J. Maharashtra Agric. Univ.* **1** (Addi.): 195-199.

Mardia K. V (1972). *Statistics and directional data*, Academic Press Ltd, London, p.357.

Martz, L. W. and Garbrecht, J. (1993). Automated extraction of drainage networks and watershed data from digital elevation models, *Wat. Resour. Bull.,* **29**: 901-908.

Matthess, G. (1982) *The Properties of Groundwater.* 406, John Wiley & Sons, New York.

McFadden, L.D. and Mc Auliffe, J.R. (1997). Lithologically influenced geomorphic responses to Holocene climatic changes in the southern Colorado plateau, Arizona: A soil-geomorphic and ecological perspective, *Geomorphology*, **19**: 303-332.

McGill, J.T. (1964). The growing importance of Urban Geology. USGS, circular, pp. 487.

Medlicott, H.B. (1865). Geological notes on Assam and the hill to the south. *Mem. Geol. Surv. India*, **4**. *Pt. 3:* 416-442.

Medlicott, H.B. (1873). The plains of the united provinces. *Rec. Geol. Surv. India,* **6**(1): 58-87.

Melton, M.A.(1957). An analysis of the relation among elements of climate, surface properties and geomorphology. Off. Nav. Res. (U.S.), Geogr. Branch, Project 389-042, Tech. Rep., **11**: 102.

Mertes, L.A.K. (1994). Rates of floodplain sedimentation on the Central Amazon River. *Geology*, **22**: 171-174.

Mertes, L.A.K., Dunne, T. and Martinelli, L.A. (1996). Channel floodplain geomorphology along the Solimos-Amazone River, *Brazil. Geol. Soc. America. Bull.,* **108** (9): 1089-1107.

Merz, B. and Bárdossy, A. (in press). Effects of spatial variability on the rainfall runoff process in a small loess catchment. *Journal of Hydrology*.

Merz, B. and Plate, E. J. (1997). An analysis of the effects of spatial variability of soil and soil moisture on runoff. *Water Resources Research*, **33**(12): 2909-2922.

Miller, V.C. (1953). A Quantitative geomorphic study of drainage basin characteristics in Clinch Mt. Area Verginea and Tennessec. Tech.. Rep. No.3, Dept. Geog. Columbia Univ., New York, Contract N6 ONR 271-030, pp. 1-30.

Miller, O.M and Summerson, C.H. (1960). Slope zone Maps. *Geographical Res.*, **50**: 194-202.

Mishra G.P. and Yadav, A.K. (1978). A comparative study of physico-chemical characteristics of river and river water in Central India. *J. Hydrobiologia,* **59**: 275-278.

Moore, I. D., Gessler, P. E., Nielsen, G. A., and Peterson, G. A. (1993). Soil attribute prediction using digital terrain analysis. *Soil Science Society of America Journal*, **57**:443-452.

Moore, I. D., Grayson, R. B., and Ladson, A. R. (1991). Digital terrain modelling: a review of hydrological, geomorphological, and biological applications. *Hydrological Processes*, **5**(3): 3-30.

Morrison, W.G.(1994). *Dictionary of geology*. CBS Pub., New Delhi, p. 200.

Mote, S.R. and Shekdar, A.V. (2002). GIS based planning of solid waste management. *'Spatial Information Technology: Remote Sensing and GIS- ICORG'* edited by I.V. Murali Krishna, **I**: 297-302.

Moussa R. (1997). Geomorphological transfer function calculated from digital elevation models for distributed hydrological modeling. *Hydrological Processes* **11**: 429–449.

MRSAC (1992). Maharashtra Remote Sensing Application Centre. Dept. of planning, Govt. of Maharashtra, Nagpur. Unpublished Geomorphological map of Parbhani district.

Mukherjee, B., Mehrotra, P.N., Pati, M., Tetarway, S., Prasad, T., Mukherjee, M. and Dhar, S. (1991). A preliminary study of the Geomorphology and Environmental Dynamics of a stream Ecosystem at Dasong in Ranchi. *J. Ecobio.* **3**(1): 43-51.

Murlidharan, D. and Athavale, R. N. (1998). Base paper on artificial recharge. National Geophysical Research Institute, p. 67.

Murthy, K.N. and Jayaram, K.M. (1996). Groundwater resources and development potential of Parbhani district, Maharashtra, C.G.W.B Report on Annual Action Programme (1995-96). 702/DIS/96, Nagpur, pp. 1-39.

Nair, K.K.K, Chatterjee, A.K. and Sano, T. (1996). Stratigraphy and geochemistry of the Deccan basalts along Toranmal section, Wn. Satpura region. *Gondwana Geol. Magz.*, Spl. **2**: 23-48.

Nakamura, F. (1986). Analysis of storage and transport processes based on age distribution of sediment. *Tans. Jpn. Geomorph. Union*, **7**: 165-184.

Nakamura, F and Kikuchi, S.I.(1996). Some methodological development in the analysis of sediment transport processes using age distribution of floodplain deposits. *Geomorphology*, **16**: 139-145.

Nathawat, M.S, Punia, M.S., Sani, R.R. and Sanjay Raj (1984). Application of remote sensing data for Geomorphological and Environmental Conservation studies. *Indian Geomorphology* (ed. by H.S.Sharma) CBC, New Delhi, Publ., pp. 307-312.

Nayak, T.R. (2002). GIS for soil erosion modeling: A case study of Bebas Watershed, Sagar District (M.P.) India. *'Spatial Information Technology: Remote Sensing and GIS- ICORG'* edited by I.V. Murali Krishna, **1**: 191-197.

NCR (National Research Council) (1977). Drinking water and health. National Research Councils, **4**: National Academy Press, Washington D.C.

NRC (National Research Council) (1993) *Situ Bioremediation: When does it work?* National Academy Press, Washington, DC.

O'Callaghan, J. F. and Mark, D. M. (1984). The extraction of drainage networks from digital elevation data, Comp. Vision Graphics Image Process., **28**: 323-344.

Ofoegbu, G. I., Bagtzoglou, A. C., Green, R. T., and Muller, M. (1999). Effects of perched water on thermally driven moisture flow at the pro-posed Yucca Mountain repository for high-level waste. *Nuclear Technology,* **125**(2): 235–253.

Ofoegbu, G. I., Painter, S., Chen, R., Fedors, R. W., and Ferrill, D. A. (2001). Geomechanical and thermal effects on moisture flow at the pro-posed Yucca Mountain nuclear waste repository. *Nuclear Technology* **134**(3): 241–262.

Oldham, R.D.(1917). Structure of Gangetic plains. Mem. Geol. Surv. India, **42**: 2.

O'Leary, D.W. Freidman, J.D., and Pohn, H.A. 1976. Lineament, linear, lineation: Some proposed new definitions for old terms. *Geological Society of America Bulletin*. **87**: 1463-1469.

Ollier, C.D. (1981). Tectonics and Land form. Longman London and New York.

O'Loughlin, E. M. 1986. Prediction of surface saturation zones in natural catchments by topographic analysis, *Water Resources Res*., **22**: 794-804.

Palanivel, S., Ganesh, A. And Vasantha Kumaran, T. (1996) Geohydrological evaluation of upper Agniar and Vellar Basins, Tamilnadu: An integrated approach using Remote sensing, Geophysical and well inventory data. *J. Indian Soc. Remote Sensing*. **24** (3): 153-168.

Pande, B.N., K. Kumar, A.K. Lal and P.K.L. Das (1993). A preliminary study on the physico-chemical quality of the water of river Koshi at Purnia. *J. Ecobiologia,* **5**(3): 237-239.

Parker, C. (1977). World-wide variations in hydraulic geometry exponents of stream channels: An analysis and some observations. *J.of Hydrology,* **33**: 133-146.

Pascoe, E.H.(1965).A manual of the geology of India and Burma. Govt. of India Press, *G.S.I,* Calcutta, 3rd edition, **1**: 1-485.

Patil, B.S., Khadilkar, A.K. and Zambre, M.K. (1999). Shallow groundwater zones mapping by using remote sensing techniques: A case study around Pishore, Aurangabad district, Maharashtra. In Siminar Vol. On *"Groundwater and Watershed Development"* at Jai Hind College, Dhule, pp. 63-65.

Patil, V.Y. (1990). Performance of bore wells in the Manjra basin in Beed Dist. *Proc. Vol. of seminar on "Modern techniques of rainwater harvesting, water conservation and artificial recharge for drinking water, afforestation, horticulture and agriculture"*, pp. 146-149.

Peshwa, V.V., Mulay, J.G. and Kale V.S. (1987). Fracture zone in the Deccan traps of western and central India: A study based on remote sensing Techniques, Photonirvachak *J. Indian Soc. Rem. Sensing ,* **15**(1),: 9-17.

Pilgrim, G.E. (1905). On the occurrence of Elephas antiquus (nomadicus) in the Godavari Alluvium. *Rec. Geol. Surv. India,* **32**: Pt. 32: 199-218.

Piper, C.S. (1966). Soil and plant analysis. Hans Publ. Bombay, pp. 89-128.

Pofali, R.M, Gaikwad, S.T. and Swaminanathan, M. (1995). Study of geomorphic features in Chandrapur district, Maharashtra. *Indian Geomorphology* (ed. by S.R.Jog), Rawat Publications, Jaipur, **II**: 221-229.

Pofali, R.M., Sehgal, J.I. and Swaminathan, M. (1985). Problems and Potentials of Pedogeomorphic surfaces in the Basin of Purna. *Indian Geomorphology* (edited by H.S.Sharma) CBC Publ. New Delhi, pp. 81-93.

Powar, K.B. (1981). Lineament fabric and Dyke patterns in the western part of the Deccan volcanic province. Mem Geol. Soc. India, **3**: 45-57.

Prasad, B.N. and S. Manjula (1986). Ecological study of blue green algae in river Gomati. *Indian J. Env. Health,* **22**(2): 151-168.

Prasanna, P.R. (2002). Cadastral level resource information system for Marapalam watershed in the Nilgiri District, South India using remote sensing, GPS and GIS. *In proceeding volume of the international symposium of ISPRS Commission VII on Resource and Environmental*

Monitoring held during December 3-6, 2002, **XXXIV** *Part 7*, pp.915-921.

Pugh, J.C. (1966). Landforms of low latitude. *Essays in Geomorphology* (ed. G.H. Dury) Heinmann, London.

Qasim, S.Z. and Siddiqui, R. H. (1960). Preliminary observations of river Kali covered by the effluent of industrial wastes. *Current Science,* **29**(5): 310-311.

Qian, N. and Dai, D. (1980). The problems of river sedimentation and the present status of its research in China, in *Proceedings of the International Symposium on River Sedimentation*, 24 29 Alarch 1980, Beijing: The Chinese Society of Hydraulic Engineering (Vol. 1), Guanghua Press (UNESCO), 19-39.

Qureshy M. N and Hinze W. J. (1989). Workshop Overview. *Proc.Joint Indo -US Workshop on Regional geophysical lineaments-their tectonic and economic significances*, Mem.Geol.Soc.Ind., Bangalore, No.12, pp.3-10.

Rai, H. (1974). Limnological studies on the river Yamuna at Delhi, India, Part II, the dynamics of phytoplankton population in the river Yamuna. *Arch. Hydrobiol.,* **73**: 492-517.

Raise, E. and Henry, J. (1937). An average slope map of Southern New England, *Geographical Review,* **27**: 467-472.

Raj, R., Maurya, D.M. and Chamyal, L.S. (1999). Evolution of Mahi drainage basin during Quaternary: A morphometric approach. *Gondwana Geol. Magz. Spl.* **4**: 131-139.

Rajaguru, S.N. (1968-69). Some new fossil discoveries from Western Maharashtra, India. *Puratattva*, **2**: 16-22.

Rajaguru, S.N. (1969).On the Late Pleistocene of the Deccan Quaternaria, **XI** Roma, 241-253.

Rajaguru, S.N. (1970). Studies in the Late Pleistocene of the Mula-Mutha valley Unpublished Ph.D. thesis, Poona Univ.

Rajaguru, S.N. and Kale, V.S. (1985). Changes in the fluvial regime of Western Maharashtra upland rivers during Late Quaternary. *J. Geol. Soc. India,* **26**, 16-27.

Rajaguru, S.N., Kale, V.S. and Badam, G.L. (1993). Quaternary fluvial systems in Upland Maharashtra. *Current Sci.* **64** (11 & 12): 817-822.

Raju, K.C.B. (1985). Artificial recharge through percolation tanks and sub-surface dykes. *In proceeding volume on Artificial Recharge of Groundwater*, January 1985, pp. 12-21.

Raju, K.C.B. (1995). Importance of recharging depleted aquifers: State of the art of artificial recharge in India. *J. Geol. Soc. India,* **54**(4): 429-454.

Raju, K.C.B., Rao, G.V.K. and Kumar, B.J. (1985). Analytical Aspects of remote sensing techniques for groundwater prospecting in hard rocks. *Proc. Sixth Asian Conference on remote sensing,* pp.127-132.

Raju, K.C.B. (1998). Importance of recharging depleted aquifer: State of artificial recharge in *India. J. Geol. Soc.,* India, **51** (4), pp.429-454.

Rajurkar, S.T., Bhate, V.D. and Sharma, S.B. (1990). Lineament fabric of Madhya Pradesh and Maharashtra and its tectonic significance. *GSI* Spl. Pub. No. 28, pp. 241-259.

Ramanuju Chary K.R. and Subbarao, N.V. (2002). Design of artificial recharge Structures to improve the groundwater quality. *In proceeding volume of the international conference on Hydrology and Watershed Management* held during December 18-20, 2002, **I**: 339-349.

Ramesh Y (1990). Geomorphic Studies in Upper Gostani River Basin with Special References to Borra (Karst) Caves Visakapatnam District, A.P., India. Ph.D. Thesis, Andhra University, India, 131p.

Rao, V.N.R., Mohan, R., Hariprasad, V and Ramasubramanian, R. (1993). Seasonal dynamics of physico-chemical factors in a tropical high altitude lake: An assessment in relation to phytoplankton. *J. Environ. Biol.,* **14**(1): 063-075.

Ratnaparkhi, T.G. (1975). Underground bunds for artificial recharge. Approaches and methodologies for development of groundwater resources, *Proc. Indo-German Workshop,* Hyderabad, pp. 353-359.

Ravishankar, T. and Venkataratnam, L. (2002). Remote sensing and GIS approach for land productivity assessment. *In proceeding volume of the international symposium of ISPRS Commission VII on Resource and Environmental Monitoring* held during December 3-6, 2002, **XXXIV** *Part 7*: 256-261.

Reddy, D.V. (1995). *Engineering geology for civil engineers*, Oxf & IBH Pub. Co. Pvt. Ltd., New Delhi.

Reddy, C.S.S., D.J. Campagna, and D.W. Levandowski. (1989). Digital image processing of multitemporal Landsat data and its application in groundwater exploration: Quantitative Remote Sensing: An Economic

Tool for the Nineties, International Geosciences & Remote Sensing Symposium, **2**: L1-L6.

Reddy, M.P. and Venkateswarulu, V. (1987). Assessment of water quality and pollution in the river Tungabhadra near Kurnool, Andhra Pradesh. *J. Environ. Biol.,* **8**(2): 109-119.

Rich, J.L. (1916). A graphical method of determining the average inclination of a land surface from a contour map. *Trans. Illinois Academy of Science,* **9**: 196-199.

Richard, K.S.(1982). General problems in morphometry. In: Geomorphological Techniques (ed. by A. Goudie), George Allen and Unwin, London, pp. 26-30.

Ritchie, J. C., Menenti, M. and Weltz, M. A. (1996). 'Measurements of land surface features using an airborne laser altimeter: the HAPEX-Sahel experiment', *Int. J. Rem. Sens.,* **17**: 3705-3724.

Ritter. D. F. and others. 2002. *Process Geomorphology* 4th ed. William C. Brown Publishers, 560 pp.

Rodriguez-Iturbe, I. and Valdes, J. B. (1979). The geomorphologic structure of hydrologic response. *Water Resources Research*, **15**(6):1409-1420.

Roy, H.K. (1955). Plankton ecology of river Hoogli at Palta, W. Bengal. Ecology, **36**: 169-175.

Roy, R.K. and Chattopadhyay, G.S. (1997). Quaternary geology of the Environs of Ganga delta, West Bengal and Bihar. *Indian J. Geology*, **69** (2): 177-209.

Saaty T.L. (1980). *The Analytic Hierarchy Process*. Mcgraw Hill, New York.

SAC, (1992) Space Applications Centre, A GIS Based Information System for Regional Planning-A case study for Bharatpur district, ISRO, Ahemdabad, July 1992.

Sanders, L. L. 1998. *A Manual of Field Hydrogeology*. Prentice Hall, Inc.

Sankar K Jegatheesan M .S and Balasubramanian A (1996). Geoelectrical Resistivity Studies in the Kanyakumari District, Tamil Nadu, *J. of Applied Hydrology* **IX**, Nos (1&2): 83-90.

Sara Naaz and Venkateshwara Rao, B. (1998). Application of Geographic Information system for the Identification of soil conservation sites in the parts of Sriramsagar catchment, *Indian proc of workshop on watershed management-problems and prospects (WMPP1998),* Centre for Water Resources, J.N. Tech. Univ., Hyderabad, pp. 43-53.

Sardhukan, M.M. (1998). An integrated approach for rainwater harvesting and water conservation technique by suitable structures within the micro-watershed in Deccan Trap terrain of Maharashtra. *Proc. Vol. Of seminar on "Modern techniques of rainwater harvesting, water conservation and artificial recharge for drinking water, afforestation, horticulture and agriculture"*, pp. 271-274.

Sarkar, A. (1995). Morphometric identity of Topologically Distinct Channel Network Basin: A case study of Lodhan Khola basin, Darjeeling Himalayas. *Indian J. Geology,* **67** (2): 142-150.

Satyanarayana , R. (1991) Remote sensing studies on the land and water resources of Hyderabad City and environs. Ph.D. Thesis, S.V. Univ. Tirupati, India (Unpublished).

Sauer, V. B., Thomas, Jr., W. O., Strickler, V., and Wilson, K. (1983). Flood characteristics of urban watersheds in the united states. U.S. Geological Survey Water-Supply Paper, 2207.

Saxena, D.N. and Sharma, S.P. (1981). Zooplankton fauna of some lantic water of Gwalior at Govind Sagar Chattri tank, Sawarkar Sarovar and Matsyasarovar. *Environment India*, **4**: 113-117.

Scheidegger, A.E. (1970. *Theoretical Geomorphology*. 2nd ed. Springerv-erlag. Berlin, p. 22.

Scheidegger,A.E.(1973). Hydrogeomorphology. *J.Hydrology,* **20**(2): 193-215.

Schmidt, J. and Dikau, R. (submitted). Extracting geomorphometric attributes and objects from digital elevation models - semantics, methods, future needs. In Dikau, R. and Saurer, H., editors, *GIS in Physical Geography*.

Schmidt, J. and Gärtner, H. (submitted). Investigations of geomorphometric significance in hydrologic processes using geo-information-technologies — results and implications for gis needs in geomorphology. *Earth Surface Processes and Landforms*.

Schmidt, J., Merz, B., and Dikau, R. (in press). Morphological structure and hydrological process modelling. *Zeitschrift für Geomorphologie, Supplement Band*.

Schumm, S.A. (1954). The relation of drainage basin relief to sediment loss. Assoc. Ind. Sci. Hydrol. Proc. Ass. Gen. Rome. **1** (36): 216-219.

Schumm, S.A. (1956). The evolution of drainage systems and slopes in bad lands at Perth, Amboi, New Jersey. *Geol. Soc. Ame. Bull.* **67** (5): 597-646.

Schumm, S.A. (1977). The fluvial systems. John Wiley, New York.

Schumm, S.A., Erskine, W.D. and Tilleard, J.W.(1996). Morphology, hydrology and evlolution of the anastomosing Ovens and King rivers, Victoria, *Australia. GSA Bull*, **108** (10): 1212-1224.

Schumm, S.A. and Mosely M.P. (1973). Slope morphology. Dowden, Hutchinson and Ross, Inc. Stroudrburg, Pennnsylvania, pp. 13-60.

Schumm, S.A. and Spitz, W.J. (1996). Geological influences on the lower Mississippi river and its alluvial valley. *Engineering Geology*, **45**: 245-261.

Seay, W.M. (1979), Southern Appalachian Tectonic Study: Tennessee Valley Authority, Division of Water Management, Geologic Services Branch, p. 66.

Sen, G and Cohen, T.H. (1994). Deccan intrustion, crustal extension, doming and the size of the Deccan Reunion Plume head. Subbarao, K.V. (ed). Volcanisam, Wiley Eastern Ltd., pp. 201-216.

Shah, P.N. (2002). Hydrogeomorphological mapping for evaluation of groundwater prospecting in Mirzapur District (U.P.) India by using IRS IA, LISS II geocoded data. *'Spatial Information Technology: Remote Sensing and GIS- ICORG'* edited by I.V. Murali Krishna, II: 447-454.

Shaikh, M. Arif, Francis, R.D. and Prabhakar, J.D. (1997). Hydrobiological studies on Sina river at Ahmednagar, Maharashtra, *India. J. Ecobiol.,* **9**(3): 207-210.

Sharma A. N. (1981). Spatial Approach for District Planning: A case study of Karnal District, Concept Publishing Company, New Delhi.

Sharma, S.B. (1984). Geomorphological study of Maharashtra and M.P. based on the landsat MSS. Division GSI Central Region, Nagpur, (Unpublished document), pp. 1-45.

Sharma, J.N and Amin, N. (1996). A morphometric study of the Dikhou river basin, India. *Indian J. Geomorphology*, **1**(2): 207-224.

Sharma D and Jugran D. K (1992). Hydrogeomorphological Studies around Pinjaur-Morni-Kala Amb Area, Ambala District (Hariyana) and Sirmour District (Himachala Pradesh). Photonirvachak, *J. Indian Society Remote Sensing,* **20**(4): 187-197.

Sharma, J.P. Raja, P., Maji, A.K. and Gajbhiye, K.S. (2002). Evaluation of soil suitability for different crops using GIS: A case study in Vardha District in Maharashtra, India. *In proceeding volume of the international symposium of ISPRS Commission VII on Resource and Environmental Monitoring* held during December 3-6, 2002, **XXXIV** *Part 7*: 256-261.

Sharma, S.K. Sharma, S.K. Mehta, M. and Manwali, S. (2000). *"Guide on artificial recharge to groundwater"*, Central Groundwater Board, Pub. New Delhi, pp. 93.

Shimer, J. A. (1972). Field Guide to Landforms in the United States. The MacMillan Company, New York. pp. 272.

Shreve, R.L. (1966). Statistical law of stream numbers. *J. Geol.* **74** (1): 17-37.

Shrivastava, R.K., Pandit, M.K. and Rajani Upadhyaya (1988). Petrography and chemical stratigraphy of Deccan basalt from a part of Malwa plateau: Use of cluster analysis in chemical stratigraphy. Geol. Soc. India, Mem. **10**: 181-198.

Shukla, S.C., Tripathi, B.D., Mishra, B.P. and Chaturvedi, S.S. (1992). Physico-chemical and bacteriological properties of water of the river Ganga at Gazipur. *Comp. Physiol. Ecol.,* **17**: 3-96.

Singh, D.F. (1992). Studies on the water quality index of some major rivers of Pune Maharashtra. *Proc. Acad. Environ. Biol.,* **I**: 61-66.

Singh P (1999), *Engineering & General Geology*, Kataria & Sons, Delhi.

Singh, D. R., and Gupta, P. N. (1982) Assessment of siltation in Tehri reservoir, *Proc. Internatl. Symp. Hydrological Aspects of Mountainous Watersheds*, Univ. Of Roorkee, Roorkee (4-6 November 1982). Mangalik Prakashan, Saharanpur, VIII-60 to VIII-66.

Singh, H.P. (1990). Scope of watershed management for improving the productivity of the Indian arid zone. Natl. Symp. On watershed Dev. And Management. Kanpur *IWRS Proc*., pp. 418-425.

Singh, N.C. and Sinha, L.C. (1981). Diurnal cycle of abiotic parameters at Ramsar well, Bhagalpur. *Comp. Physiol. Ecol.,* **6**(1): 38-40.

Singh, R.K. and Anandh, H. (1996). Water quality index of some Indian rivers. *Indian J. Environ. Health,* **38**(1): 21-34.

Singh, V.P. and Jain, (2002). Hydrology and watershed management education in 21st Century. *Proceed volume of the international conference on Hydrology and Watershed Management* held during December 18-20, 2002, **I**: 1-14.

Singh, R.P. (1969). Geomorphological evolution of Chhota Nagpur highland. *Natl. Geogr.* Soc. India, B.H.U. Varanasi, Res. Publ. No. 5, p.22.

Singh, R.P.(1983). The landscape cycles of Peninsular India, in contributions to Indian Geography. Geomorphology, (edited by K.R.Dikshit), Heritage Publishers, New Delhi, **2**: 87-104.

Singh, S. and Sinha, J. (2001). Application of GPS in geological mapping: A preliminary thought, *Proc. The Asian GPS conl.*, New Delhi, pp. 145-147.

Singh, Savindra (1972). Altimetric analysis: a morphometric technique of landform study. *Natl. Geographer*, **7**: 59-68.

Singh Savindra. (1978). Physiographic regions, landforms and erosion surfaces of Ranchi plateau. *Natl. Geographer*, **13**(1): 43-65.

Singh Savindra (1981). Estimation of drainage density. *Natl. Geographer*, **16**(2): 81-89.

Singh Savindra (1985). An approach to the study of landform. In concepts and methods in Geography. Geog. Dept. Burdwan Univ., pp. 38-50.

Singh Savindra (1998). *Geomorphology*, (edited by Savindra Singh), Prayag Pustak Bhavan, Allahabad, pp.334-412.

Singh Savindra, Ojha, S.S. and Agnihotri, S.P. (1984). On the regional variations in bifurcation ratios. In: Geomorphological Mosaic, edited by S.C.Mukhopadhyay, Calcutta Univ., pp. 203-210.

Singh Savindra and Rastogi, N. (1992). Morphological anc cultural regionalization of Belan-Son inter stream region of the eastern Rewa Plateau, *Natl. Geographer*, **27** (2): 159-167.

Singh Savindra and Rastogi, N. (1994). The study of morphodynamics and morphogenesis in sub humid tropical environment. *Natl. Geographer*, **29** (1): 61-72.

Singh Savindra and Srivastava, R. (1974). A morphometric study of the tributary basin of upper reaches of the Belan river. *Natl. Geographer*, **9**: 31-44.

Singh Surendra (1995). Geomorphology in the appraisal of natural resources. *Indian Geomorphology*, (ed. by S.R.Jog), Rawat Publ. Jaipur, pp. 193-210.

Sinha , B.K., Kumar Ashok, Srivastava, D.and Srivastava, S.K. (1990). Integrated approach for demarcating the fracture zone for well sit location-A case study near Gumla and Lohardaga, Bihar. *J. Indian Soc. Remote Sensing*. **18**(3): 1-8.

Small, R.J. (1970). The study of landforms. Cambridge University Press.

Smith, C.H.(1938). The morphometry of landscape. An analysis of slope. *Annals of Assoc. Am. Geogr.* **29**: 34.

Smith, K.G. (1950). Standards for grading texture of erosional topography. *Am. J. Sci.* **248**: 655-665.

Smith, R. D.; Ammann, A.; Bartoldus, C., and Brinson, M. M. (1995) Approach for Assessing Wetland Functions Using Hydrogeomorphic Classification, Reference Wetlands, and Functional Indices. Final Report. Vicksburg, MS: Army Engineer Waterways Experiment Station; WES/TR/WRP-DE-9. 90 p.

Som, S.K., Joshi, R., Roy, P.K. and Mukherjee, M.M. (1998). Morphotectonic evolution of the laterite profiles over Sukinda ultramafics, Jaipur district, Rajasthan *J. Geol. Soc. Ind.* **52**: 449-456.

Spark, B.W., (1960). *Geomorphology*. Longmans, Green, London.

Sridhar K. T., Mahadevan S., Jere A., Mohan S. and Mohan A. (1996). Siting of Beach Resorts: A GIS Based Study, GEOMATICS 96: Data Standardization, Ahemdabad, April 1996.

Sridhar K. T., Mahadevan S. and Mukund Rao (1994). Relational Database for Geographic Applications, COMMAD 94, Bangalore, December 1994.

Srinivasa Rao, Y., Krishna Raddy, T.V. and Nayudu, P.T. (1997). Hydrogeomorphological studies by remote sensing Application in Niva River basin, Chittor District, Andhra Pradesh. Photonirvachak *J. Indian Soc. Rem. Sensing*, **25**(3): 187-194.

Starks, P. J. and Humes, K. S. (1996). Hydrology Data Report: Washita '94. NAWQL 96-1. US Department of Agriculture-Agricultural Research Service, Durant, Oklahoma.

Stoddart, D.R. (1965). Climatic geomorphology: Review and re-assessment. *Progress in Physical Geography*, **1**: 159-222.

Stoddart, D.R. (1969). In Water, Earth and Man (ed. Chorley, R.J.), Methuen, London, pp. 43.

Strahler, A.N. (1952). Hypsometric (area-altitude) analysis of erosional topography. *Geol. Soc. Am. Bull.* **63** (11): 1117-1142.

Strahler, A.N. (1954). Statistical analysis in geomorphic research. *J. Geol.* **62:** 1-25.

Strahler, A.N. (1956). Quantitative slope analysis. *Geol. Soc. Am. Bull.* **67** (5): 571-596.

Strahler, A.N. (1957). Quantitative analysis of watershed geomorphology. *Tans. Am. Geophys. Union*, **38**: 931.

Strahler, A.N. (1964). Quantitative geomorphology of drainage basin and channel networks. *Handbook of Applied Hydrology* (edited by V.T.Chow), pp. 4.39-476.

Strahler, A.N. (1968). Quantitative geomorphology. In: Fairbridge, R.W. (ed). The Encylcopedia of geomorphology, Reinhold Book Crop. New York.

Subbarao, K.V., Bodas, M.S., Hooper, P.R. and Walsh, J.N. (1988). Petrogenesis of Jawahar and Igatpuri formations, Western Deccan Basalt province. *Geol. Soc. Ind.,* Mem. **10**: 253-280.

Subbarao, K.V., Chandrashekharam, D., Navneethakrishna, P. and Hooper, P.R. (1994). Stratigraphy and structure of parts of Central Deccan Basalt Province: Erruptive models. In: Subbarao, K.V. (eds.), Volcanism, Wiley Eastern Ltd., pp. 321-332.

Subramanyan, S.K and Mani, G. (1981). Genetical geomorphic aspects of laterites on high and low landforms in parts of Tamil Nadu (India). In: Laterisation Process, A.A. Balkemia, Rotteralam, pp. 236-245.

Subramanyan, V. (1974). A quantitative analysis of two drainage basins around Sagar, M.P. *Indian J. Geol.* **40**(1): 76-99.

Sulebak, J. R., Etzelmüller, B. and Sollid, J. L.(1997). Landscape regionalization by automatic classification of landform elements. *Norsk Geografisk Tidsskrift*, **51**(1): 35-45.

Suvarna, A.C. and Somashekar, R.K. (1997). Ecological studies on the riverine ecosystem of Karnataka: The physico-chemical characteristics of river Cauvery. *J. Environment and Pollution,* **4**(3): 57-63.

Tarboton, D. G. (1997). A new method for the determination of flow directions and upslope areas in grid digital elevation models', *Water Resour. Res.,* **33**: 309-319.

Tarboton, D. G., Bras, R. L. and Rodiriguez-Iturbe, I. 1991. On the extraction of channel networks from digital elevation data, *Hydrol. Process.*, 5: 81-100.

Taylor, L.E. (1981). Environmental aspects of water resources planning in England and Waler. *J. of Hydrology,* **51**: 231-243.

Tejwani, K. G. (1986). 'Sedimentation of reservoirs in the Himalayan region and estimates of river load', unpub. report prepared for Mohonk Mountain Conference, April 1986, Boulder, Colorado: International Mountain Society (unpublished).

Theodossiou, E. I. and Dowman, I. J. (1990). Heighting accuracy of SPOT, *Photogramm. Engg Rem. Sens.*, **56**: 1643-1649.

Thorat, P.K. and Sable, A.B. (1990). Importance of the Geological Mapping in selection of sites for artificial recharge in the Deccan Volcanics with reference to some case histories from Parmer Taluka of Ahmednagar Dist., Maharashtra. In Seminar Vol. On *"Modern Techniques of rainwater harvesting, water conservation and artificial drinking water afforestation, horticulture and agriculture"*, pp. 199-239.

Tiwari, M.P. (1999). Quaternary lighostratigraphic formations of Central Indian river basin, their correlation and chronology. *Gondwana Geol. Magz. Spl.* **4**: 17-31.

Tiwari, M.P. and Bhai, H.Y. (1997a). Geomorphology of the alluvial fill of the Narmada valley. Geol. Surv. India., Spl. Pub. No. 46, pp. 9-18.

Tiwari, M.P. and Bhai, H.Y. (1997b). Quaternary stratigraphy of the Narmada valley. . Geol. Surv. India., Spl. Pub. No. 46, pp. 33-63.

Tiwari, A and Rai, B. (1996). Hydrogeomorphological mapping for groundwater prospecting using Landsat MSS images- A case study of part of Dhanbad district. *J. Indian Soc. Remote Sensing,* **24**(4): 281-285.

Todd, D.K. (1980). *Groundwater Hydrology.* 2nd edition, John Wiley, New York, pp. 535.

Todd, D.K. (1995). *Groundwater hydrology*. John Wiley. New York, pp. 282.

Toutin, T. (1998). SPOT and Landsat stereo fusion for data extraction over mountainous areas, *Photogramm. Engg Rem. Sens.*, **64**: 109-113.

Toutin, T. and Beaudoin, M. (1995). Real-time extraction of planimetric and altimetric features from digital stereo SPOT data using a digital video plotter, *Photogramm. Engg Rem. Sens.*, ***61***: 63-68.

Trivedi, R.K. and Goel, P.K. (1986). Chemical and Biological Methods for Water Pollution Studies (R.K. Trivedi and P.K. Goel ed.) Environmental Publ., Karad, India, pp. 35-80.

Tucker, (1979). Red and photographic infrared linear combinations for monitoring vegetation: *Remote Sensing of Environment* **8**: 127-150.

Twidale, C.R. (1997). Some recently developed landforms: Climatic implications. *Geomorphology*, **19**: 349-365.

UNEP (United Nations Environment Programme) (1989) *Environmental Data Report 1989/90,* Blackwell Reference, Oxford, 547 pp.

UNEP-IETC (1999). Technical Publication Series 11, Planning and Management of lakes and reservoirs: An integrated approach to eutrophication.

UNESCO (2000). Report on the status of the IHP VI draft plan.

USDA 1981. WRC Hydrologic Unit Boundaries Correlated with Oklahoma CNI, Watersheds, and Other Sub-state Resource Unit Delineations, Report no. 6-81. Soil Conservation Service, Stillwater, Oklahoma.

Vaidhyanadhan, R. (1983). Landform studies of the Peninsular India. A survey of research in physical geography, Concept Publ. Co., New Delhi, pp. 47-60.

Vazir Mahamood and Durga Rao, K.H.V. (2002). Groundwater modeling using remote sensing and GIS: A case study of Visakhapatnam (A.P.) India. *'Spatial Information Technology: Remote Sensing and GIS- ICORG'* edited by I.V. Murali Krishna, **I**: 162-168.

Verma, S.R., Tyagi, A.K. and Dalela, R.C. (1978). Physico-chemical and biological characteristics of Kadrabad Drain in U.P. *Indian J. Environ.,* **20**(1): 1-13.

Vladimirescu, I.(1978). Hydrology. Ed. Didectica si pedagogica, Bucharest (in Romanian), quoted from Zavoianu, I. (1985). pp. 185.

Ward, R.C. (1978). Q_{max} or MAF? – Some reflections on hydrology today. *Geography*, **63**(4): 301- 313.

Weaver, G.D. (1965). What is landform? *Professional Geographers*, **XVII**, No. 1: 11-13.

Wegner, M.J. (1999). Making crop sense with GPS. *Geo-Europe*, **4**: 30-32.

Welch, E.B. (1998). Ecological effects of wastewater. Publ. by the press syndicate of the University of Cambridge, 337 pp.

Wentworth, C.K. (1930). A simplified method of determining the average slope of land surface *Am. J. Science.* **20**: 184 -194.

Whalley, W.B. (1976). Properties of materials and Geomorphological Explanations. Whalley, W.B. (ed.) Oxford Univ. Press.

White, W.B., (1999) Conceptual Models for Karstic Aquifers; in Karst Waters Institute Special Publication 5, Proceedings of the symposium held February 24 through 27, Charlottesville, VA, pp. 11-16.

WHO (1984). World Health Organization. International standards for drinking Water Tech. Rep., 70 pp.

WHO (1992). Revision of WHO guidelines for drinking water quality: Report of the final task group meeting at Geneva Switzerland, pp. 21-25.

Wittwer, C. S., Chen, G., and Bodvarsson, G. S. (1993). Studies of the role of fault zones in fluid flow using the site-scale numerical model of Yucca Mountain. *Fourth International High Level Radioactive Waste Management Conference*. Las Grange Park, IL: American Nuclear Society, pp. 667–674.

Wood, W.F. and Snell, J.B. (1960). A quantitative system for classifying landforms. United States Army, Quarter Master Res. Centre, Natick, Massachastts, Tech. Rep. Ep.124, 20 p.

Wood, E. F. (1995). Heterogeneity and scaling land-atmospheric water and energy fluxes in climate systems. In Feddes, R. A., editor, *Space and time scale variability and interdepencies in hydrological processes*, pp. 3-19. Cambridge university Press.

Wood, E. F., Sivapalan, M., and Beven, K. (1990). Similarity and scale in catchment storm response. *Reviews of Geophysics*, **28**(1):1-18.

Woodruff, C.M., Jr., Caran, S.C., Gever, C., Henry, C.D., Macperson, G.L., and McBride, M.W., (1982), Geothermal Resource Assessment for the State of Texas: Bureau of Economic Geology, University of Texas, Austin, Texas.

Woodruff, C.M., Jr., and Caran, S.C. (1984). Lineaments of Texas—expressions of surface and subsurface features: Technical Papers, American Congress and Mapping, American Society of Photogammetry Fall Convention, San Antonio, pp. 741-749.

Worsley, P. (1979). Whither geomorphology? *Area*, **11**: 97-101.

Yedekar, D.B., Aramaki, S., Fujii, T and Sano, T. (1996). Geochemical signature and stratigraphy of the Chhindwara-Jabalpur-Seoni-Mandla sector of the eastern Deccan Volcanic Province and Problems of its correlation. *Gondwana Geol. Magz*. Spl. **2**: 49-68.

Zavoianu, I. (1985). Morphometry of drainage basins. Developments in water science, Zo, Elsevier, Amsterdam (Ed Ion Zavoianu), 63-195.

Zernitz, E.R. (1932). Drainage patterns and their significance. *J. Geol.* **40**: 498-521.

Zhang, D.D. (1998). Geomorphological problems of the middle reaches of the Tsangpo river, Tibet. *Earth Surface Processes and Landforms*, **23** (10): 889-903.

Index

D

E

F

G

H

I

J

K

L

M

N

O

P

Q

R

FSC
www.fsc.org
MIX
Papier | Fördert gute Waldnutzung
FSC® C083411

Zeitfracht Medien GmbH
Ferdinand-Jühlke-Straße 7
99095 Erfurt, Deutschland
produktsicherheit@kolibri360.de